Information Technology and Organizational Learning

Managing Behavioral Change in the Digital Age

Third Edition

Information Technology and Organizational Learning

Managing Behavioral Change in the Digital Age

Third Edition

Arthur M. Langer

CRC Press
Taylor & Francis Group
Boca Raton London New York

CRC Press is an imprint of the
Taylor & Francis Group, an **informa** business

Contents

Foreword

Digital technologies are transforming the global economy. Increasingly, firms and other organizations are assessing their opportunities, developing and delivering products and services, and interacting with customers and other stakeholders digitally. Established companies recognize that digital technologies can help them operate their businesses with greater speed and lower costs and, in many cases, offer their customers opportunities to co-design and co-produce products and services. Many start-up companies use digital technologies to develop new products and business models that disrupt the present way of doing business, taking customers away from firms that cannot change and adapt. In recent years, digital technology and new business models have disrupted one industry after another, and these developments are rapidly transforming how people communicate, learn, and work.

Against this backdrop, the third edition of Arthur Langer's *Information Technology and Organizational Learning* is most welcome. For decades, Langer has been studying how firms adapt to new or changing conditions by increasing their ability to incorporate and use advanced information technologies. Most organizations do not adopt new technology easily or readily. Organizational inertia and embedded legacy systems are powerful forces working against the adoption of new technology, even when the advantages of improved technology are recognized. Investing in new technology is costly, and it requires aligning technology with business strategies and transforming corporate cultures so that organization members use the technology to become more productive.

Information Technology and Organizational Learning addresses these important issues—and much more. There are four features of the new edition that I would like to draw attention to that, I believe, make this a valuable book. First, Langer adopts a behavioral perspective rather than a technical perspective. Instead of simply offering normative advice about technology adoption, he shows how sound learning theory and principles can be used to incorporate technology into the organization. His discussion ranges across the dynamic learning organization, knowledge management, change management, communities of practice, and virtual teams. Second, he shows how an organization can move beyond technology alignment to true technology integration. Part of this process involves redefining the traditional support role of the IT department to a leadership role in which IT helps to drive business strategy through a technology-based learning organization. Third, the book contains case studies that make the material come alive. The book begins with a comprehensive real-life case that sets the stage for the issues to be resolved, and smaller case illustrations are sprinkled throughout the chapters, to make concepts and techniques easily understandable. Lastly, Langer has a wealth of experience that he brings to his book. He spent more than 25 years as an IT consultant and is the founder of the Center for Technology Management at Columbia University, where he directs certificate and executive programs on various aspects of technology innovation and management. He has organized a vast professional network of technology executives whose companies serve as learning laboratories for his students and research. When you read the book, the knowledge and insight gained from these experiences is readily apparent.

If you are an IT professional, *Information Technology and Organizational Learning* should be required reading. However, anyone who is part of a firm or agency that wants to capitalize on the opportunities provided by digital technology will benefit from reading the book.

Charles C. Snow

Professor Emeritus, Penn State University

Co-Editor, Journal of Organization Design

Acknowledgments

Many colleagues and clients have provided significant support during the development of the third edition of *Information Technology and Organizational Learning*.

I owe much to my colleagues at Teachers College, namely, Professor Victoria Marsick and Lyle Yorks, who guided me on many of the theories on organizational learning, and Professor Lee Knefelkamp, for her ongoing mentorship on adult learning and developmental theories. Professor David Thomas from the Harvard Business School also provided valuable direction on the complex issues surrounding diversity, and its importance in workforce development.

I appreciate the corporate executives who agreed to participate in the studies that allowed me to apply learning theories to actual organizational practices. Stephen McDermott from ICAP provided invaluable input on how chief executive officers (CEOs) can successfully learn to manage emerging technologies. Dana Deasy, now global chief information officer (CIO) of JP Morgan Chase, contributed enormous information on how corporate CIOs can integrate technology into business strategy. Lynn O'Connor Vos, CEO of Grey Healthcare, also showed me how technology can produce direct monetary returns, especially when the CEO is actively involved.

And, of course, thank you to my wonderful students at Columbia University. They continue to be at the core of my inspiration and love for writing, teaching, and scholarly research.

Author

Arthur M. Langer, EdD, is professor of professional practice of management and the director of the Center for Technology Management at Columbia University. He is the academic director of the Executive Masters of Science program in Technology Management, vice chair of faculty and executive advisor to the dean at the School of Professional Studies and is on the faculty of the Department of Organization and Leadership at the Graduate School of Education (Teachers College). He has also served as a member of the Columbia University Faculty Senate. Dr. Langer is the author of *Guide to Software Development: Designing & Managing the Life Cycle.* 2nd Edition (2016), *Strategic IT: Best Practices for Managers and Executives* (2013 with Lyle Yorks), *Information Technology and Organizational Learning* (2011), *Analysis and Design of Information Systems* (2007), *Applied Ecommerce* (2002), and *The Art of Analysis* (1997), and has numerous published articles and papers, relating to digital transformation, service learning for underserved populations, IT organizational integration, mentoring, and staff development. Dr. Langer consults with corporations and universities on information technology, cyber security, staff development, management transformation, and curriculum development around the Globe. Dr. Langer is also the chairman and founder of Workforce Opportunity Services (www.wforce.org), a non-profit social venture that provides scholarships and careers to underserved populations around the world.

Dr. Langer earned a BA in computer science, an MBA in accounting/finance, and a Doctorate of Education from Columbia University.

Introduction

Background

Information technology (IT) has become a more significant part of workplace operations, and as a result, information systems personnel are key to the success of corporate enterprises, especially with the recent effects of the digital revolution on every aspect of business and social life (Bradley & Nolan, 1998; Langer, 1997, 2011; Lipman-Blumen, 1996). This digital revolution is defined as a form of "disruption." Indeed, the big question facing many enterprises today is, How can executives anticipate the unexpected threats brought on by technological advances that could devastate their business? This book focuses on the vital role that information and digital technology organizations need to play in the course of organizational development and learning, and on the growing need to integrate technology fully into the processes of workplace organizational learning. Technology personnel have long been criticized for their inability to function as part of the business, and they are often seen as a group outside the corporate norm (Schein, 1992). This is a problem of cultural assimilation, and it represents one of the two major fronts that organizations now face in their efforts to gain a grip on the new, growing power of technology, and to be competitive in a global world. The other major front concerns the strategic integration of new digital technologies into business line management.

Because technology continues to change at such a rapid pace, the ability of organizations to operate within a new paradigm of dynamic change emphasizes the need to employ action learning as a way to build competitive learning organizations in the twenty-first century. *Information Technology and Organizational Learning* integrates some of the fundamental issues bearing on IT today with concepts from organizational learning theory, providing comprehensive guidance, based on real-life business experiences and concrete research.

This book also focuses on another aspect of what IT can mean to an organization. IT represents a broadening dimension of business life that affects everything we do inside an organization. This new reality is shaped by the increasing and irreversible dissemination of technology. To maximize the usefulness of its encroaching presence in everyday business affairs, organizations will require an optimal understanding of how to integrate technology into everything they do. To this end, this book seeks to break new ground on how to approach and conceptualize this salient issue—that is, that the optimization of information and digital technologies is best pursued with a synchronous implementation of organizational learning concepts. Furthermore, these concepts cannot be implemented without utilizing theories of strategic learning. Therefore, this book takes the position that technology literacy requires individual and group strategic learning if it is to transform a business into a technology-based learning organization. *Technology-based organizations* are defined as those that have implemented a means of successfully integrating technology into their process of organizational learning. Such organizations recognize and experience the reality of technology as part of their everyday business function. It is what many organizations are calling "being digital."

This book will also examine some of the many existing organizational learning theories, and the historical problems that have occurred with companies that have used them, or that have failed to use them. Thus, the introduction of technology into organizations actually provides an opportunity to reassess and reapply many of the past concepts, theories, and practices that have been used to support the importance of organizational learning. It is important, however, not to confuse this message with a reason for promoting organizational learning, but rather, to understand the seamless nature of the relationship between IT and organizational learning. Each needs the other to succeed. Indeed, technology has only served to expose problems that have existed in organizations for decades, e.g., the inability to drive down responsibilities to the operational levels of the organization, and to be more agile with their consumers.

This book is designed to help businesses and individual managers understand and cope with the many issues involved in developing organizational learning programs, and in integrating an important component: their IT and digital organizations. It aims to provide a combination of research case studies, together with existing theories on organizational learning in the workplace. The goal is also to provide researchers and corporate practitioners with a book that allows them to incorporate a growing IT infrastructure with their existing workforce culture. Professional organizations need to integrate IT into their organizational processes to compete effectively in the technology-driven business climate of today. This book responds to the complex and various dilemmas faced by many human resource managers and corporate executives regarding how to actually deal with many marginalized technology personnel who somehow always operate outside the normal flow of the core business.

While the history of IT, as a marginalized organization, is relatively short, in comparison to that of other professions, the problems of IT have been consistent since its insertion into business organizations in the early 1960s. Indeed, while technology has changed, the position and valuation of IT have continued to challenge how executives manage it, account for it, and, most important, ultimately value its contributions to the organization. Technology personnel continue to be criticized for their inability to function as part of the business, and they are often seen as outside the business norm. IT employees are frequently stereotyped as "techies," and are segregated in such a way that they become isolated from the organization. This book provides a method for integrating IT, and redefining its role in organizations, especially as a partner in formulating and implementing key business strategies that are crucial for the survival of many companies in the new digital age. Rather than provide a long and extensive list of common issues, I have decided it best to uncover the challenges of IT integration and performance through the case study approach.

IT continues to be one of the most important yet least understood departments in an organization. It has also become one of the most significant components for competing in the global markets of today. IT is now an integral part of the way companies become successful, and is now being referred to as the digital arm of the business. This is true across all industries. The role of IT has grown enormously in companies throughout the world, and it has a mission to provide strategic solutions that can make companies more competitive. Indeed, the success of IT, and its ability to operate as part of the learning organization, can mean the difference between the success and failure of entire companies. However, IT must be careful that it is not seen as just a factory of support personnel, and does not lose its justification as driving competitive advantage. We see in many organizations that other digital-based departments are being created, due to frustration with the traditional IT culture, or because they simply do not see IT as meeting the current needs for operating in a digital economy.

This book provides answers to other important questions that have challenged many organizations for decades. First, how can managers master emerging digital technologies, sustain a relationship with organizational learning, and link it to strategy and performance? Second, what is the process by which to determine the value of using technology, and how does it relate to traditional ways of calculating return on investment, and establishing risk models? Third, what are the cyber security implications of technology-based products and services? Fourth, what are the roles and responsibilities of the IT executive, and the department in general? To answer these questions, managers need to focus on the following objectives:

- Address the operational weaknesses in organizations, in terms of how to deal with new technologies, and how to better realize business benefits.
- Provide a mechanism that both enables organizations to deal with accelerated change caused by technological innovations, and integrates them into a new cycle of processing, and handling of change.
- Provide a strategic learning framework, by which every new technology variable adds to organizational knowledge and can develop a risk and security culture.

- Establish an integrated approach that ties technology accountability to other measurable outcomes, using organizational learning techniques and theories.

To realize these objectives, organizations must be able to

- create dynamic internal processes that can deal, on a daily basis, with understanding the potential fit of new technologies and their overall value within the structure of the business;
- provide the discourse to bridge the gaps between IT- and non-IT-related investments, and uses, into one integrated system;
- monitor investments and determine modifications to the life cycle;
- implement various organizational learning practices, including learning organization, knowledge management, change management, and communities of practice, all of which help foster strategic thinking, and learning, and can be linked to performance (Gephardt & Marsick, 2003).

The strengths of this book are that it integrates theory and practice and provides answers to the four common questions mentioned. Many of the answers provided in these pages are founded on theory and research and are supported by practical experience. Thus, evidence of the performance of the theories is presented via case studies, which are designed to assist the readers in determining how such theories and proven practices can be applied to their specific organization.

A common theme in this book involves three important terms: *dynamic*, *unpredictable*, and *acceleration*. *Dynamic* is a term that represents spontaneous and vibrant things—a motive force. Technology behaves with such a force and requires organizations to deal with its capabilities. Glasmeier (1997) postulates that technology evolution, innovation, and change are dynamic processes. The force then is technology, and it carries many motives, as we shall see throughout this book. *Unpredictable* suggests that we cannot plan what will happen or will be needed. Many organizational individuals, including executives, have attempted to predict when, how, or why technology will affect their organization. Throughout our recent history, especially during the "digital disruption" era, we have found that it is difficult, if not impossible, to predict how technology will ultimately benefit or hurt organizational growth and competitive advantage. I believe that technology is volatile and erratic at times. Indeed, harnessing technology is not at all an exact science; certainly not in the ways in which it can and should be used in today's modern organization. Finally, I use the term *acceleration* to convey the way technology is speeding up our lives. Not only have emerging technologies created this unpredictable environment of change, but they also continue to change it rapidly—even from the demise of the dot-com era decades ago. Thus, what becomes important is the need to respond quickly to technology. The inability to be responsive to change brought about by technological innovations can result in significant competitive disadvantages for organizations.

This new edition shows why this is a fact especially when examining the shrinking S-Curve. So, we look at these three words—dynamic, unpredictable, and acceleration—as a way to define how technology affects organizations; that is, technology is an accelerating motive force that occurs irregularly. These words name the challenges that organizations need to address if they are to manage technological innovations and integrate them with business strategy and competitive advantage. It only makes sense that the challenge of integrating technology into business requires us first to understand its potential impact, determine how it occurs, and see what is likely to follow. There are no quick remedies to dealing with emerging technologies, just common practices and sustained processes that must be adopted for organizations to survive in the future.

I had four goals in mind in writing this book. First, I am interested in writing about the challenges of using digital technologies strategically. What particularly concerns me is the lack of literature that truly addresses this issue. What is also troublesome is the lack of reliable techniques for the evaluation of IT, especially since IT is used in almost every aspect of business life. So, as we increase our use and dependency on technology, we seem to understand less about how to measure and validate its outcomes. I also want to convey my thoughts about the importance of embracing nonmonetary methods for evaluating technology, particularly as they relate to determining return on investment. Indeed, indirect and nonmonetary benefits need to be part of the process of assessing and approving IT projects.

Second, I want to apply organizational learning theory to the field of IT and use proven learning models to help transform IT staff into becoming better members of their organizations. Everyone seems to know about the inability of IT people to integrate with other departments, yet no one has really created a solution to the problem. I find that organizational learning techniques are an effective way of coaching IT staff to operate more consistently with the goals of the businesses that they support.

Third, I want to present cogent theories about IT and organizational learning; theories that establish new ways for organizations to adapt new technologies. I want to share my experiences and those of other professionals who have found approaches that can provide positive outcomes from technology investments.

Fourth, I have decided to express my concerns about the validity and reliability of organizational learning theories and practices as they apply to the field of IT. I find that most of these models need to be enhanced to better fit the unique aspects of the digital age. These modified models enable the original learning techniques to address IT-specific issues. In this way, the organization can develop a more holistic approach toward a common goal for using technology.

Certainly, the balance of how technology ties in with strategy is essential. However, there has been much debate over whether technology should drive business strategy or vice versa. We will find that the answer to this is "yes." Yes, in the sense that technology can affect the way organizations determine their missions and business strategies; but "no" in that technology should not be the only component for determining mission and strategy. Many managers have realized that business is still business, meaning that technology is not a "silver bullet." The challenge, then, is to determine how best to fit technology into the process of creating and supporting business strategy. Few would doubt today that technology is, indeed, the most significant variable affecting business strategy. However, the most viable approach is to incorporate technology into the *process* of determining business strategy. I have found that many businesses still formulate their strategies first, and then look at technology, as a means to efficiently implement objectives and goals. Executives need to better understand the unique and important role that technology provides us; it can drive business strategy, and support it, at the same time.

Managers should not solely focus their attention on generating breakthrough innovations that will create spectacular results. Most good uses of technology are much subtler, and longer-lasting. For this reason, this book discusses and defines new technology life cycles that blend business strategy and strategic learning. Building on this theme, I introduce the idea of *responsive organizational dynamism* as the core theory of this book. Responsive organizational dynamism defines an environment that can respond to the three important terms (dynamic, unpredictable, and acceleration). Indeed, technology requires organizations that can sustain a system, in which individuals can deal with dynamic, unpredictable, and accelerated change, as part of their regular process of production. The basis of this concept is that organizations must create and sustain such an environment to be competitive in a global technologically-driven economy. I further analyze responsive organizational dynamism in its two subcomponents: strategic integration and cultural assimilation, which address how technology needs to be measured as it relates to business strategy, and what related social–structural changes are needed, respectively.

Change is an important principle of this book. I talk about the importance of how to change, how to manage such change, and why emerging technologies are a significant agent of change. I support the need for change, as an opportunity to use many of the learning theories that have been historically difficult to implement. That is, implementing change brought on by technological innovation is an opportunity to make the organization more "change ready" or, as we define it today, more "agile." However, we also know that little is known about how organizations should actually go about modifying existing processes to adapt to new technologies and become digital entities—and to be accustomed to doing this regularly. Managing through such periods of change requires that we develop a model that can deal with dynamic, unpredictable, and accelerated change. This is what responsive organizational dynamism is designed to do.

We know that over 20% of IT projects still fail to be completed. Another 54% fail to meet their projected completion date. We now sit at the forefront of another technological spurt of innovations that will necessitate major renovations to existing legacy systems, requiring that they be linked to sophisticated e-business systems. These e-business systems will continue to utilize the Internet, and emerging mobile technologies. While we tend to focus primarily on what technology generically does, organizations need urgently to prepare themselves for the next generation of advances, by forming structures that can deal with continued, accelerated change, as the norm of daily operations. For this edition, I have added new sections and chapters that address the digital transformation, ways of dealing with changing consumer behavior, the need to form evolving cyber security cultures, and the importance of integrating Gen Y employees to accelerate competitive advantage.

This book provides answers to a number of dilemmas but ultimately offers an imbricate cure for the problem of latency in performance and quality afflicting many technologically-based projects. Traditionally, management has attempted to improve IT performance by increasing technical skills and project manager expertise through new processes. While there has been an effort to educate IT managers to become more interested and participative in business issues, their involvement continues to be based more on service than on strategy. Yet, at the heart of the issue is the entirety of the organization. It is my belief that many of the programmatic efforts conducted in traditional ways and attempting to mature and integrate IT with the rest of the organization will continue to deliver disappointing results.

My personal experience goes well beyond research; it draws from living and breathing the IT experience for the past 35 years, and from an understanding of the dynamics of what occurs inside and outside the IT department in most organizations. With such experience, I can offer a path that engages the participation of the entire management team and operations staff of the organization. While my vision for this kind of digital transformation is different from other approaches, it is consistent with organizational learning theories that promote the integration of individuals, communities, and senior management to participate in more democratic and visionary forms of thinking, reflection, and learning. It is my belief that many of the dilemmas presented by IT have existed in other parts of organizations for years, and that the Internet revolution only served to expose them. If we believe this to be true, then we must begin the process of integrating technology into strategic thinking and stop depending on IT to provide magical answers, and inappropriate expectations of performance.

Technology is not the responsibility of any one person or department; rather, it is part of the responsibility of every employee. Thus, the challenge is to allow organizations to understand how to modify their processes, and the roles and responsibilities of their employees, to incorporate digital technologies as part of normal workplace activities. Technology then becomes more a subject and a component of discourse. IT staff members need to emerge as specialists who participate in decision making, development, and sustained support of business evolution. There are also technology-based topics that do not require the typical expertise that IT personnel provide. This is a literacy issue that requires different ways of thinking and learning during the everyday part of operations. For example, using desktop tools, communicating via e-mail, and saving files and data, are integral to everyday operations. These activities affect projects, yet they are not really part of the responsibilities of IT departments. Given the knowledge that technology is everywhere, we must change the approach that we take to be successful. Another way of looking at this phenomenon is to define technology more as a commodity, readily available to all individuals. This means that the notion of technology as organizationally segregated into separate cubes of expertise is problematic, particularly on a global front.

Thus, the overall aim of this book is to promote organizational learning that disseminates the uses of technology throughout a business, so that IT departments are a partner in its use, as opposed to being its sole owner. The cure to IT project failure, then, is to engage the business in technology decisions in such a way that individuals and business units are fundamentally involved in the process. Such processes need to be designed to dynamically respond to technology opportunities and thus should not be overly bureaucratic. There is a balance between establishing organizations that can readily deal with technology versus those that become too complex and inefficient.

This balance can only be attained using organizational learning techniques as the method to grow and reach technology maturation.

Overview of the Chapters

Chapter 1 provides an important case study of the Ravell Corporation (a pseudonym), where I was retained for over five years. During this period, I applied numerous organizational learning methods toward the integration of the IT department with the rest of the organization. The chapter allows readers to understand how the theories of organizational learning can be applied in actual practice, and how those theories are particularly beneficial to the IT community. The chapter also shows the practical side of how learning techniques can be linked to measurable outcomes, and ultimately related to business strategy. This concept will become the basis of integrating learning with strategy (i.e., "strategic learning"). The Ravell case study also sets the tone of what I call the IT dilemma, which represents the core problem faced by organizations today. Furthermore, the Ravell case study becomes the cornerstone example throughout the book and is used to relate many of the theories of learning and their practical applicability in organizations. The Ravell case has also been updated in this second edition to include recent results that support the importance of alignment with the human resources department.

Chapter 2 presents the details of the IT dilemma. This chapter addresses issues such as isolation of IT staff, which results in their marginalization from the rest of the organization. I explain that while executives want technology to be an important part of business strategy, few understand how to accomplish it. In general, I show that individuals have a lack of knowledge about how technology and business strategy can, and should, be linked, to form common business objectives. The chapter provides the results of a three-year study of how chief executives link the role of technology with business strategy. The study captures information relating to how chief executives perceive the role of IT, how they manage it, and use it strategically, and the way they measure IT performance and activities.

Chapter 3 focuses on defining how organizations need to respond to the challenges posed by technology. I analyze technological dynamism in its core components so that readers understand the different facets that comprise its many applications. I begin by presenting technology as a *dynamic variable* that is capable of affecting organizations in a unique way. I specifically emphasize the unpredictability of technology, and its capacity to accelerate change—ultimately concluding that technology, as an independent variable, has a dynamic effect on organizational development. This chapter also introduces my theory of responsive organizational dynamism, defined as a disposition in organizational behavior that can respond to the demands of technology as a dynamic variable. I establish two core components of responsive organizational dynamism: *strategic integration* and *cultural assimilation*. Each of these components is designed to tackle a specific problem introduced by technology. Strategic integration addresses the way in which organizations determine how to use technology as part of business strategy. Cultural assimilation, on the other hand, seeks to answer how the organization, both structurally and culturally, will accommodate the actual human resources of an IT staff and department within the process of implementing new technologies. Thus, strategic integration will require organizational changes in terms of cultural assimilation. The chapter also provides a perspective of the technology life cycle so that readers can see how responsive organizational dynamism is applied, on an IT project basis. Finally, I define the driver and supporter functions of IT and how these contribute to managing technology life cycles.

Chapter 4 introduces theories on organizational learning, and applies them specifically to responsive organizational dynamism. I emphasize that organizational learning must result in individual, and organizational transformation, that leads to measurable performance outcomes. The chapter defines a number of organizational learning theories, such as reflective practices, learning organization, communities of practice, learning preferences and experiential learning, social discourse, and the use of language. These techniques and approaches to promoting organizational learning are then configured into various models that can be used to assess individual and organizational development. Two important models are designed to be used in responsive organizational dynamism: the applied individual learning wheel and the technology maturity arc. These models lay the foundation for my position that learning maturation involves a steady linear progression from an individual focus toward a system or organizational perspective. The chapter also addresses implementation issues—political challenges that can get in the way of successful application of the learning theories.

Chapter 5 explores the role of management in creating and sustaining responsive organizational dynamism. I define the tiers of middle management in relation to various theories of management participation in organizational learning. The complex issues of whether organizational learning needs to be managed from the top down, bottom up, or middle-top-down are discussed and applied to a model that operates in responsive organizational dynamism. This chapter takes into account the common three-tier structure in which most organizations operate: executive, middle, and operations. The executive level includes the chief executive officer (CEO), president, and senior vice presidents. The middle is the most complex, ranging from vice president/director to supervisory roles. Operations covers what is commonly known as "staff," including clerical functions. The knowledge that I convey suggests that all of these tiers need to participate in management, including operations personnel, via a self-development model. The chapter also presents the notion that knowledge management is necessary to optimize competitive advantage, particularly as it involves transforming tacit knowledge into explicit knowledge. I view the existing theories on knowledge management, create a hybrid model that embraces technology issues, and map them to responsive organizational dynamism. Discussions on change management are included as a method of addressing the unique ways that technology affects product development. Essentially, I tie together responsive organizational dynamism with organizational change theory, by offering modifications to generally accepted theories. There is also a specific model created for IT organizations, that maps onto organizational-level concepts. Although I have used technology as the basis for the need for responsive organizational dynamism, I show that the needs for its existence can be attributed to any variable that requires dynamic change. As such, I suggest that readers begin to think about the next "technology" or variable that can cause the same needs to occur inside organizations. The chapter has been extended to address the impact of social networking and the leadership opportunities it provides to technology executives.

Chapter 6 examines how organizational transformation occurs. The primary focus of the chapter is to integrate transformation theory with responsive organizational dynamism. The position taken is that organizational learning techniques must inevitably result in organizational transformation. Discussions on transformation are often addressed at organizational level, as opposed to focusing on individual development. As in other sections of the book, I extend a number of theories so that they can operate under the auspices of responsive organizational dynamism, specifically, the works of Yorks and Marsick (2000) and Aldrich (2001). I expand organizational transformation to include ongoing assessment within technology deliverables. This is accomplished through the use of a modified Balanced Scorecard originally developed by Kaplan and Norton (2001). The Balanced Scorecard becomes the vehicle for establishing a strategy-focused and technology-based organization.

Chapter 7 deals with the many business transformation projects that require outsource arrangements and virtual team management. This chapter provides an understanding of when and how to consider outsourcing and the intricacies of considerations once operating with virtual teams. I cover such issues as management considerations and the challenges of dealing in multiple locations. The chapter extends the models discussed in previous chapters so that they can be aligned with operating in a virtual team environment. Specifically, this includes communities of practice, social discourse, self-development, knowledge management, and, of course, responsive organizational dynamism and its corresponding maturity arcs. Furthermore, I expand the conversation to include IT and non-IT personnel, and the arguments for the further support needed to integrate all functions across the organization.

Chapter 8 presents updated case studies that demonstrate how my organizational learning techniques are actually applied in practice. Three case studies are presented: Siemens AG, ICAP, and HTC. Siemens AG is a diverse international company with 20 discrete businesses in over 190 countries. The case study offers a perspective of how a corporate chief information officer (CIO) introduced e-business strategy. ICAP is a leading international money and security broker. This case study follows the activities of the electronic trading community (ETC) entity, and how the CEO transformed the organization and used organizational learning methods to improve competitive advantage. HTC (a pseudonym) provides an example of why the chief IT executive should report to the CEO, and how a CEO can champion specific projects to help transform organizational norms and behaviors. This case study also maps the transformation of the company to actual examples of strategic advantage.

Chapter 9 focuses on the challenges of forming a "cyber security" culture. The growing challenges of protecting companies from outside attacks have established the need to create a cyber security culture. This chapter addresses the ways in which information technology organizations must further integrate with business operations, so that their firms are better equipped to protect against outside threats. Since the general consensus is that no system can be 100% protected, and that most system compromises occur as a result of internal exposures, information technology leaders must educate employees on best practices to limit cyberattacks. Furthermore, while prevention is the objective, organizations must be internally prepared to deal with attacks and thus have processes in place should a system become penetrated by third-party agents.

Chapter 10 explores the effects of the digital global economy on the ways in which organizations need to respond to the consumerization of products and services. From this perspective, digital transformation involves a type of social reengineering that affects the ways in which organizations communicate internally, and how they consider restructuring departments. Digital transformation also affects the risks that organizations must take in what has become an accelerated changing consumer market.

Chapter 11 provides conclusions and focuses on Gen Y employees who are known as "digital natives" and represent the new supply chain of talent. Gen Y employees possess the attributes to assist companies to transform their workforce to meet the accelerated change in the competitive landscape. Most executives across industries recognize that digital technologies are the most powerful variable to maintaining and expanding company markets. Gen Y employees provide a natural fit for dealing with emerging digital technologies. However, success with integrating Gen Y employees is contingent upon Baby Boomer and Gen X management adopting new leadership philosophies and procedures suited to meet the expectations and needs of these new workers. Ignoring the unique needs of Gen Y employees will likely result in an incongruent organization that suffers high turnover of young employees who will ultimately seek a more entrepreneurial environment.

Chapter 12 seeks to define best practices to implement and sustain responsive organizational dynamism. The chapter sets forth a model that creates separate, yet linked, best practices and maturity arcs that can be used to assess stages of the learning development of the chief IT executive, the CEO, and the middle management. I discuss the concept of *common threads*, by which each best practices arc links through common objectives and outcomes to the responsive organizational dynamism maturity arc presented in Chapter 4. Thus, these arcs represent an integrated and hierarchical view of how each component of the organization contributes to overall best practices. A new section has been added that links ethics to technology leadership and maturity.

Chapter 13 summarizes the many aspects of how IT and organizational learning operate together to support the responsive organizational dynamism environment. The chapter emphasizes the specific key themes developed in the book, such as evolution versus revolution; control and empowerment; driver and supporter operations; and responsive organizational dynamism and self-generating organizations. Finally, I provide an overarching framework for "organizing" reflection and integrate it with the best practices arcs.

As a final note, I need to clarify my use of the words *information technology, digital technology,* and *technology*. In many parts of the book, they are used interchangeably, although there is a defined difference. Of course, not all technology is related to information or digital; some is based on machinery or the like. For the purposes of this book, the reader should assume that IT and digital technology are the primary variables that I am addressing. However, the theories and processes that I offer can be scaled to all types of technological innovation.

1

THE "RAVELL" CORPORATION

Introduction

Launching into an explanation of information technology (IT), organizational learning, and the practical relationship into which I propose to bring them is a challenging topic to undertake. I choose, therefore, to begin this discussion by presenting an actual case study that exemplifies many key issues pertaining to organizational learning, and how it can be used to improve the performance of an IT department. Specifically, this chapter summarizes a case study of the IT department at the Ravell Corporation (a pseudonym) in New York City. I was retained as a consultant at the company to improve the performance of the department and to solve a mounting political problem involving IT and its relation to other departments. The case offers an example of how the growth of a company as a "learning organization"—one in which employees are constantly learning during the normal workday (Argyris, 1993; Watkins & Marsick, 1993)—utilized reflective practices to help it achieve the practical strategic goals it sought. Individuals in learning organizations integrate processes of learning into their work. Therefore, a learning organization must advocate a system that allows its employees to interact, ask questions, and provide insight to the business. The learning organization will ultimately promote systematic thinking, and the building of organizational memory (Watkins & Marsick, 1993). A learning organization (discussed more fully in Chapter 4) is a component of the larger topic of organizational learning.

The Ravell Corporation is a firm with over 500 employees who, over the years, had become dependent on the use of technology to run its business. Its IT department, like that of many other companies, was isolated from the rest of the business and was regarded as a peripheral entity whose purpose was simply to provide technical support. This was accompanied by actual physical isolation—IT was placed in a contained and secure location away from mainstream operations. As a result, IT staff rarely engaged in active discourse with other staff members unless specific meetings were called relating to a particular project. The Ravell IT department, therefore, was not part of the community of organizational learning—it did not have the opportunity to learn along with the rest of the organization, and it was never asked to provide guidance in matters of general relevance to the business as a whole. This marginalized status resulted in an us-versus-them attitude on the part of IT and non-IT personnel alike.

Much has been written about the negative impact of marginalization on individuals who are part of communities. Schlossberg (1989) researched adults in various settings and how marginalization affected their work and self-efficacy. Her theory on marginalization and mattering is applied to this case study because of its relevance and similarity to her prior research. For example, IT represents similar characteristics to a separate group on a college campus or in a workplace environment. Its physical isolation can also be related to how marginalized groups move away from the majority population and function without contact. The IT director, in particular, had cultivated an adversarial relationship with his peers. The director had shaped a department that fueled his view of separation. This had the effect of further marginalizing the position of IT within the organization. Hand in hand with this form of separatism came a sense of actual dislike on the part of IT personnel for other employees. IT staff members were quick to point fingers at others and were often noncommunicative with members of other departments within the organization. As a result of this kind of behavior, many departments lost confidence in the ability of IT to provide support; indeed, the quality of support that IT furnished had begun to deteriorate. Many departments at Ravell began to hire their own IT support personnel and were determined to create their own information systems subdepartments. This situation eventually became unacceptable to management, and the IT director was terminated. An initiative was begun to refocus the department and its position within the organization. I was retained to bring about this change and to act as the IT director until a structural transformation of the department was complete.

A New Approach

My mandate at Ravell was initially unclear—I was to "fix" the problem; the specific solution was left up to me to design and implement. My goal became one of finding a way to integrate IT fully into the organizational culture at Ravell. Without such integration, IT would remain isolated, and no amount of "fixing" around this issue would address the persistence of what was, as well, a cultural problem. Unless IT became a true part of the organization as a whole, the entire IT staff could be replaced without any real change having occurred from the organization's perspective. That is, just replacing the entire IT staff was an acceptable solution to senior management. The fact that this was acceptable suggested to me that the knowledge and value contained in the IT department did not exist or was misunderstood by the senior management of the firm. In my opinion, just eliminating a marginalized group was not a solution because I expected that such knowledge and value did exist, and that it needed to be investigated properly. Thus, I rejected management's option and began to formulate a plan to better understand the contributions that could be made by the IT department. The challenge was threefold: to improve the work quality of the IT department (a matter of performance), to help the department begin to feel itself a part of the organization as a whole and vice versa (a matter of cultural assimilation), and to persuade the rest of the organization to accept the IT staff as equals who could contribute to the overall direction and growth of the organization (a fundamental matter of strategic integration).

My first step was to gather information. On my assignment to the position of IT director, I quickly arranged a meeting with the IT department to determine the status and attitudes of its personnel. The IT staff meeting included the chief financial officer (CFO), to whom IT reported. At this meeting, I explained the reasons behind the changes occurring in IT management. Few questions were asked; as a result, I immediately began scheduling individual meetings with each of the IT employees. These employees varied in terms of their position within the corporate hierarchy, in terms of salary, and in terms of technical expertise. The purpose of the private meetings was to allow IT staff members to speak openly, and to enable me to hear their concerns. I drew on the principles of action science, pioneered by Argyris and Schön (1996), designed to promote individual self-reflection regarding behavior patterns, and to encourage a productive exchange among individuals. Action science encompasses a range of methods to help individuals learn how to be reflective about their actions. By reflecting, individuals can better understand the outcomes of their actions and, especially, how they are seen by others. This was an important approach because I felt learning had to start at the individual level as opposed to attempting group learning activities. It was my hope that the discussions I orchestrated would lead the IT staff to a better understanding than they had previously shown, not only of the learning process itself, but also of the significance of that process. I pursued these objectives by guiding them to detect problem areas in their work and to undertake a joint effort to correct them (Argyris, 1993; Arnett, 1992).

Important components of reflective learning are single-loop and double-loop learning. Single-loop learning requires individuals to reflect on a prior action or habit that needs to be changed in the future but does not require individuals to change their operational procedures with regard to values and norms. Double-loop learning, on the other hand, does require both change in behavior and change in operational procedures. For example, people who engage in double-loop learning may need to adjust how they perform their job, as opposed to just the way they communicate with others, or, as Argyris and Schön (1996, p. 22) state, "the correction of error requires inquiry through which organizational values and norms themselves are modified."

Despite my efforts and intentions, not all of the exchanges were destined to be successful. Many of the IT staff members felt that the IT director had been forced out, and that there was consequently no support for the IT function in the organization. There was also clear evidence of internal political division within the IT department; members openly criticized each other. Still other interviews resulted in little communication. This initial response from IT staff was disappointing, and I must admit I began to doubt whether these learning methods would be an antidote for the department. Replacing people began to seem more attractive, and I now understood why many managers prefer to replace staff, as opposed to investing in their transformation. However, I also knew that learning is a gradual process and that it would take time and trust to see results.

I realized that the task ahead called for nothing short of a total cultural transformation of the IT organization at Ravell. Members of the IT staff had to become flexible and open if they were to become more trusting of one another and more reflective as a group (Garvin, 2000; Schein, 1992). Furthermore, they had to have an awareness of their history, and they had to be willing to institute a vision of partnering with the user community. An important part of the process for me was to accept the fact that the IT staff were not habitually inclined to be reflective. My goal then was to create an environment that would foster reflective learning, which would in turn enable a change in individual and organizational values and norms (Senge, 1990).

The Blueprint for Integration

Based on information drawn from the interviews, I developed a preliminary plan to begin to integrate IT into the day-to-day operations at Ravell, and to bring IT personnel into regular contact with other staff members. According to Senge (1990), the most productive learning occurs when skills are combined in the activities of advocacy and inquiry. My hope was to encourage both among the staff at Ravell. The plan for integration and assimilation involved assigning IT resources to each department; that is, following the logic of the self-dissemination of technology, each department would have its own dedicated IT person to support it. However, just assigning a person was not enough, so I added the commitment to actually relocate an IT person into each physical area. This way, rather than clustering together in an area of their own, IT people would be embedded throughout the organization, getting first-hand exposure to what other departments did, and learning how to make an immediate contribution to the productivity of these departments. The on-site IT person in each department would have the opportunity to observe problems when they arose—and hence, to seek ways to prevent them—and, significantly, to share in the sense of accomplishment when things went well. To reinforce their commitment to their respective areas, I specified that IT personnel were to report not only to me but also to the line manager in their respective departments. In addition, these line managers were to have input on the evaluation of IT staff. I saw that making IT staff officially accountable to the departments they worked with was a tangible way to raise their level of commitment to the organization. I hoped that putting line managers in a supervisory position, would help build a sense of teamwork between IT and non-IT personnel. Ultimately, the focus of this approach was to foster the creation of a tolerant and supportive cultural climate for IT within the various departments; an important corollary goal here was also to allow reflective reviews of performance to flourish (Garvin, 1993).

Enlisting Support

Support for this plan had to be mustered quickly if I was to create an environment of trust. I had to reestablish the need for the IT function within the company, show that it was critical for the company's business operations, and show that its integration posed a unique challenge to the company. However, it was not enough just for me to claim this. I also had to enlist key managers to claim it. Indeed, employees will cooperate only if they believe that self-assessment and critical thinking are valued by management (Garvin, 2000). I decided to embark on a process of arranging meetings with specific line managers in the organization. I selected individuals who would represent the day-to-day management of the key departments. If I could get their commitment to work with IT, I felt it could provide the stimulus we needed. Some line managers were initially suspicious of the effort because of their prior experiences with IT. However, they generally liked the idea of integration and assimilation that was presented to them, and agreed to support it, at least on a trial basis.

Predictably, the IT staff were less enthusiastic about the idea. Many of them felt threatened, fearing that they were about to lose their independence or lose the mutual support that comes from being in a cohesive group. I had hoped that holding a series of meetings would help me gain support for the restructuring concept. I had to be careful to ensure that the staff members would feel that they also had an opportunity to develop a plan, that they were confident would work. During a number of group sessions, we discussed various scenarios of how such a plan might work. I emphasized the concepts of integration and assimilation, and that a program of their implementation would be experimental. Without realizing it, I had engaged IT staff members in a process of self-governance. Thus, I empowered them to feel comfortable with voicing new ideas, without being concerned that they might be openly criticized by me if I did not agree. This process also encouraged individuals to begin thinking more as a group. Indeed, by directing the practice of constructive criticism among the IT staff, I had hoped to elicit a higher degree of reflective action among the group and to show them that they had the ability to learn from one another as well as the ability to design their own roles in the organization (Argyris, 1993). Their acceptance of physical integration and, hence, cultural assimilation became a necessary condition for the ability of the IT group, to engage in greater reflective behavior (Argyris & Schön, 1996).

Assessing Progress

The next issue concerned individual feedback. How was I to let each person know how he or she was doing? I decided first, to get feedback from the larger organizational community. This was accomplished by meeting with the line managers and obtaining whatever feedback was available from them. I was surprised at the large quantity of information they were willing to offer. The line managers were not shy about participating, and their input allowed me to complete two objectives: (1) to understand how the IT staff was being perceived in its new assignment and (2) to create a social and reflective relationship between IT individuals and the line managers. The latter objective was significant, for if we were to be successful, the line managers would have to assist us in the effort to integrate and assimilate IT functions within their community.

After the discussions with managers were completed, individual meetings were held with each IT staff member to discuss the feedback. I chose not to attribute the feedback to specific line managers but rather to address particular issues by conveying the general consensus about them. Mixed feelings were also disclosed by the IT staff. After conveying the information, I listened attentively to the responses of IT staff members. Not surprisingly, many of them responded to the feedback negatively and defensively. Some, for example, felt that many technology users were unreasonable in their expectations of IT. It was important for me as facilitator not to find blame among them, particularly if I was to be a participant in the learning organization (Argyris & Schön, 1996).

Resistance in the Ranks

Any major organizational transformation is bound to elicit resistance from some employees. The initiative at Ravell proved to be no exception. Employees are not always sincere, and some individuals will engage in political behavior that can be detrimental to any organizational learning effort. Simply put, they are not interested in participating, or, as Marsick (1998) states, "It would be naïve to expect that everyone is willing to play on an even field (i.e., fairly)." Early in the process, the IT department became concerned that its members spent much of their time trying to figure out how best to position themselves for the future instead of attending to matters at hand. I heard from other employees that the IT staff felt that they would live through my tenure; that is, just survive until a permanent IT director was hired. It became difficult at times to elicit the truth from some members of the IT staff. These individuals would skirt around issues and deny making statements that were reported by other employees rather than confront problems head on. Some IT staff members would criticize me in front of other groups and use the criticism as proof that the plan for a general integration was bound to fail. I realized in a most tangible sense that pursuing change through reflective practice does not come without resistance, and that this resistance needs to be factored into the planning of any such organizationally transformative initiative.

Line Management to the Rescue

At the time that we were still working through the resistance within IT, the plan to establish a relationship with line management began to work. A number of events occurred that allowed me to be directly involved in helping certain groups solve their IT problems. Word spread quickly that there was a new direction in IT that could be trusted. Line management support is critical for success in such transformational situations. First, line management is typically comprised of people from the ranks of supervisors and middle managers, who are responsible for the daily operations of their department. Assuming they do their jobs, senior management will cater to their needs and listen to their feedback. The line management of any organization, necessarily engaged to some degree in the process of learning (a "learning organization"), is key to its staff. Specifically, line managers are responsible for operations personnel; at the same time, they must answer to senior management. Thus, they understand both executive and operations perspectives of the business (Garvin, 2000). They are often former staff members themselves and usually have a high level of technical knowledge. Upper management, while important for financial support, has little effect at the day-to-day level, yet this is the level at which the critical work of integration and the building of a single learning community must be done.

Interestingly, the line management organization had previously had no shortage of IT-related problems. Many of these line managers had been committed to developing their own IT staffs; however, they quickly realized that the exercise was beyond their expertise, and that they needed guidance and leadership. Their participation in IT staff meetings had begun to foster a new trust in the IT department, and they began to see the possibilities of working closely with IT to solve their problems. Their support began to turn toward what Watkins and Marsick (1993, p. 117) call "creating alignment by placing the vision in the hands of autonomous, cross-functional synergetic teams." The combination of IT and non-IT teams began to foster a synergy among the communities, which established new ideas about how best to use technology.

IT Begins to Reflect

Although it was initially difficult for some staff members to accept, they soon realized that providing feedback opened the door to the process of self-reflection within IT. We undertook a number of exercises, to help IT personnel understand how non-IT personnel perceived them, and how their own behavior may have contributed to these perceptions. To foster self-reflection, I adopted a technique developed by Argyris called "the left-hand column." In this technique, individuals use the right-hand column of a piece of paper to transcribe dialogues that they felt had not resulted in effective communication. In the left-hand column of the same page, participants are to write what they were really thinking at the time of the dialogue but did not say. This exercise is designed to reveal underlying assumptions that speakers may not be aware of during their exchanges and that may be impeding their communication with others by giving others a wrong impression. The exercise was extremely useful in helping IT personnel understand how others in the organization perceived them.

Most important, the development of reflective skills, according to Schön (1983), starts with an individual's ability to recognize "leaps of abstraction"—the unconscious and often inaccurate generalizations people make about others based on incomplete information. In the case of Ravell, such generalizations were deeply entrenched among its various personnel sectors. Managers tended to assume that IT staffers were "just techies," and that they therefore held fundamentally different values and had little interest in the organization as a whole. For their part, the IT personnel were quick to assume that non-IT people did not understand or appreciate the work they did. Exposing these "leaps of abstraction" was key to removing the roadblocks that prevented Ravell from functioning as an integrated learning organization.

Defining an Identity for Information Technology

It was now time to start the process of publicly defining the identity of IT. Who were we, and what was our purpose? Prior to this time, IT had no explicit mission. Instead, its members had worked on an *ad hoc* basis, putting out fires and never fully feeling that their work had contributed to the growth or development of the organization as a whole. This sense of isolation made it difficult for IT members to begin to reflect on what their mission should or could be. I organized a series of meetings to begin exploring the question of a mission, and I offered support by sharing exemplary IT mission statements that were being implemented in other organizations. The focus of the meetings was not on convincing them to accept any particular idea but rather to facilitate a reflective exercise with a group that was undertaking such a task for the first time (Senge, 1990).

The identity that emerged for the IT department at Ravell was different from the one implicit in their past role. Our new mission would be to provide technical support and technical direction to the organization. Of necessity, IT personnel would remain specialists, but they were to be specialists who could provide guidance to other departments in addition to helping them solve and prevent problems. As they became more intimately familiar with what different departments did—and how these departments contributed to the organization as a whole—IT professionals would be able to make better informed recommendations. The vision was that IT people would grow from being staff who fixed things into team members who offered their expertise to help shape the strategic direction of the organization and, in the process, participate fully in organizational growth and learning.

To begin to bring this vision to life, I invited line managers to attend our meetings. I had several goals in mind with this invitation. Of course, I wanted to increase contact between IT and non-IT people; beyond this, I wanted to give IT staff an incentive to change by making them feel a part of the organization as a whole. I also got a commitment from IT staff that we would not cover up our problems during the sessions, but would deal with all issues with trust and honesty. I also believed that the line managers would reciprocate and allow us to attend their staff meetings. A number of IT individuals were concerned that my approach would only further expose our problems with regard to quality performance, but the group as a whole felt compelled to stick with the beliefs that honesty would always prevail over politics. Having gained insight into how the rest of the organization perceived them, IT staff members had to learn how to deal with disagreement and how to build consensus to move an agenda forward. Only then could reflection and action be intimately intertwined so that after-the-fact reviews could be replaced with periods of learning and doing (Garvin, 2000).

The meetings were constructive, not only in terms of content issues handled in the discussions, but also in terms of the number of line managers who attended them. Their attendance sent a strong message that the IT function was important to them, and that they understood that they also had to participate in the new direction that IT was taking. The sessions also served as a vehicle to demonstrate how IT could become socially assimilated within all the functions of the community while maintaining its own identity.

The meetings were also designed as a venue for group members to be critical of themselves. The initial meetings were not successful in this regard; at first, IT staff members spent more time blaming others than reflecting on their own behaviors and attitudes. These sessions were difficult in that I would have to raise unpopular questions and ask whether the staff had truly "looked in the mirror" concerning some of the problems at hand. For example, one IT employee found it difficult to understand why a manager from another department was angry about the time it took to get a problem resolved with his computer. The problem had been identified and fixed within an hour, a time frame that most IT professionals would consider very responsive. As we looked into the reasons why the manager could have been justified in his anger, it emerged that the manager had a tight deadline to meet. In this situation, being without his computer for an hour was a serious problem.

Although under normal circumstances a response time of one hour is good, the IT employee had failed to ask about the manager's particular circumstance. On reflection, the IT employee realized that putting himself in the position of the people he was trying to support would enable him to do his job better. In this particular instance, had the IT employee only understood the position of the manager, there were alternative ways of resolving the problem that could have been implemented much more quickly.

Implementing the Integration: A Move toward Trust and Reflection

As communication became more open, a certain synergy began to develop in the IT organization. Specifically, there was a palpable rise in the level of cooperation and agreement, with regard to the overall goals set during these meetings. This is not to suggest that there were no disagreements but rather that discussions tended to be more constructive in helping the group realize its objective of providing outstanding technology support to the organization. The IT staff also felt freer to be self-reflective by openly discussing their ideas and their mistakes. The involvement of the departmental line managers also gave IT staff members the support they needed to carry out the change. Slowly, there developed a shift in behavior in which the objectives of the group sharpened its focus on the transformation of the department, on its acknowledgment of successes and failures, and on acquiring new knowledge, to advance the integration of IT into the core business units.

Around this time, an event presented itself that I felt would allow the IT department to establish its new credibility and authority to the other departments: the physical move of the organization to a new location. The move was to be a major event, not only because it represented the relocation of over 500 people and the technological infrastructure they used on a day-to-day basis, but also because the move was to include the transition of the media communications systems of the company, to digital technology. The move required tremendous technological work, and the organization decided to perform a "technology acceleration," meaning that new technology would be introduced more quickly because of the opportunity presented by the move. The entire moving process was to take a year, and I was immediately summoned to work with the other departments in determining the best plan to accomplish the transition.

For me, the move became an emblematic event for the IT group at Ravell. It would provide the means by which to test the creation of, and the transitioning into, a learning organization. It was also to provide a catalyst for the complete integration and assimilation of IT into the organization as a whole. The move represented the introduction of unfamiliar processes in which "conscious reflection is … necessary if lessons are to be learned" (Garvin, 2000, p. 100). I temporarily reorganized IT employees into "SWAT" teams (subgroups formed to deal with defined problems in high-pressure environments), so that they could be eminently consumed in the needs of their community partners. Dealing with many crisis situations helped the IT department change the existing culture by showing users how to better deal with technology issues in their everyday work environment. Indeed, because of the importance of technology in the new location, the core business had an opportunity to embrace our knowledge and to learn from us.

The move presented new challenges every day, and demanded openness and flexibility from everyone. Some problems required that IT listen intently to understand and meet the needs of its community partners. Other situations put IT in the role of teaching; assessing needs and explaining to other departments what was technically possible, and then helping them to work out compromises based on technical limitations. Suggestions for IT improvement began to come from all parts of the organization. Ideas from others were embraced by IT, demonstrating that employees throughout the organization were learning together. IT staff behaved assertively and without fear of failure, suggesting that, perhaps for the first time, their role had extended beyond that of fixing what was broken to one of helping to guide the organization forward into the future. Indeed, the move established the kind of "special problem" that provided an opportunity for growth in personal awareness through reflection (Moon, 1999).

The move had proved an ideal laboratory for implementing the IT integration and assimilation plan. It provided real and important opportunities for IT to work hand in hand with other departments—all focusing on shared goals. The move fostered tremendous camaraderie within the organization and became an excellent catalyst for teaching reflective behavior. It was, if you will, an ideal project in which to show how reflection in action can allow an entire organization to share in the successful attainment of a common goal. Because it was a unique event, everyone—IT and non-IT personnel alike—made mistakes, but this time, there was virtually no finger-pointing. People accepted responsibility collectively and cooperated in finding solutions. When the company recommenced operations from its new location—on time and according to schedule—no single group could claim credit for the success; it was universally recognized that success had been the result of an integrated effort.

Key Lessons

The experience of the reorganization of the IT department at Ravell can teach us some key lessons with respect to the cultural transformation and change of marginalized technical departments, generally.

Defining Reflection and Learning for an Organization

IT personnel tend to view learning as a vocational event. They generally look to increase their own "technical" knowledge by attending special training sessions and programs. However, as Kegan (1998) reminds us, there must be more: "Training is really insufficient as a sole diet of education—it is, in reality a subset of education." True education involves transformation, and transformation, according to Kegan, is the willingness to take risks, to "get out of the bedroom of our comfortable world." In my work at Ravell, I tried to augment this "diet" by embarking on a project that delivered both vocational training and education through reflection. Each IT staff person was given one week of technical training per year to provide vocational development. But beyond this, I instituted weekly learning sessions in which IT personnel would meet without me and produce a weekly memo of "reflection." The goal of this practice was to promote dialogue, in the hope that IT would develop a way to deal with its fears and mistakes on its own. Without knowing it, I had begun the process of creating a discursive community in which social interactions could act as instigators of reflective behavior leading to change.

Working toward a Clear Goal

The presence of clearly defined, measurable, short-term objectives can greatly accelerate the process of developing a "learning organization" through reflective practice. At Ravell, the move into new physical quarters provided a common organizational goal toward which all participants could work. This goal fostered cooperation among IT and non-IT employees and provided an incentive for everyone to work and, consequently, learn together. Like an athletic team before an important game, or even an army before battle, the IT staff at Ravell rallied around a cause and were able to use reflective practices to help meet their goals. The move also represented what has been termed an "eye-opening event," one that can trigger a better understanding of a culture whose differences challenge one's presuppositions (Mezirow, 1990). It is important to note, though, that while the move accelerated the development of the learning organization as such, the move itself would not have been enough to guarantee the successes that followed it. Simply setting a deadline is no substitute for undergoing the kind of transformation necessary for a consummately reflective process. Only as the culmination of a process of analysis, socialization, and trust building, can an event like this speed the growth of a learning organization.

Commitment to Quality

Apart from the social challenges it faced in merging into the core business, the IT group also had problems with the quality of its output. Often, work was not performed in a professional manner. IT organizations often suffer from an inability to deliver on schedule, and Ravell was no exception. The first step in addressing the quality problem, was to develop IT's awareness of the importance of the problem, not only in my estimation but in that of the entire company. The IT staff needed to understand how technology affected the day-to-day operations of the entire company. One way to start the dialogue on quality is to first initiate one about failures. If something was late, for instance, I asked why. Rather than addressing the problems from a destructive perspective (Argyris & Schön, 1996; Schein, 1992; Senge, 1990), the focus was on encouraging IT personnel to understand the impact of their actions—or lack of action—on the company. Through self-reflection and recognition of their important role in the organization, the IT staff became more motivated than before to perform higher quality work.

Teaching Staff "Not to Know"

One of the most important factors that developed out of the process of integrating IT was the willingness of the IT staff "not to know." The phenomenology of "not knowing" or "knowing less" became the facilitator of listening; that is, by listening, we as individuals are better able to reflect. This sense of not knowing also "allows the individual to learn an important lesson: the acceptance of what is, without our attempts to control, manipulate, or judge" (Halifax, 1999, p. 177). The IT staff improved their learning abilities by suggesting and adopting new solutions to problems. An example of this was the creation of a two-shift help desk that provided user support during both day and evening. The learning process allowed IT to contribute new ideas to the community. More important, their contributions did not dramatically change the community; instead, they created gradual adjustments that led to the growth of a new hybrid culture. The key to this new culture was its ability to share ideas, accept error as a reality (Marsick, 1998), and admit to knowing less (Halifax, 1999).

Transformation of Culture

Cultural changes are often slow to develop, and they occur in small intervals. Furthermore, small cultural changes may even go unnoticed or may be attributed to factors other than their actual causes. This raises the issue of the importance of cultural awareness and our ability to measure individual and group performance. The history of the IT problems at Ravell made it easy for me to make management aware of what we were newly attempting to accomplish and of our reasons for creating dialogues about our successes and failures. Measurement and evaluation of IT performance are challenging because of the intricacies involved in determining what represents success. I feel that one form of measurement can be found in the behavioral patterns of an organization. When it came time for employee evaluations, reviews were held with each IT staff member. Discussions at evaluation reviews focused on the individuals' perceptions of their role, and how they felt about their job as a whole. The feedback from these review meetings suggested that the IT staff had become more devoted, and more willing to reflect on their role in the organization, and, generally, seemed happier at their jobs than ever before. Interestingly, and significantly, they also appeared to be having fun at their jobs. This happiness propagated into the community and influenced other supporting departments to create similar infrastructures that could reproduce our type of successes. This interest was made evident by frequent inquiries I received from other departments about how the transformation of IT was accomplished, and how it might be translated to create similar changes in staff behavior elsewhere in the company. I also noticed that there were fewer complaints and a renewed ability for the staff to work with our consultants.

Alignment with Administrative Departments

Ravell provided an excellent lesson about the penalties of not aligning properly with other strategic and operational partners in a firm. Sometimes, we become insistent on forcing change, especially when placed in positions that afford a manager power—the power to get results quickly and through force. The example of Ravell teaches us that an approach of power will not ultimately accomplish transformation of the organization. While senior management can authorize and mandate change, change usually occurs much more slowly than they wish, if it occurs at all. The management ranks can still push back and cause problems, if not sooner, then later. While I aligned with the line units, I failed to align with important operational partners, particularly human resources (HR). HR in my mind at that time was impeding my ability to accomplish change. I was frustrated and determined to get things done by pushing my agenda. This approach worked early on, but I later discovered that the HR management was bitter and devoted to stopping my efforts. The problems I encountered at Ravell are not unusual for IT organizations. The historical issues that affect the relationship between HR and IT are as follows:

- IT has unusual staff roles and job descriptions that can be inconsistent with the rest of the organization.
- IT tends to have complex working hours and needs.
- IT has unique career paths that do not "fit" with HR standards.
- IT salary structures shift more dynamically and are very sensitive to market conditions.
- IT tends to operate in silos.

The challenge, then, to overcome these impediments requires IT to

- reduce silos and IT staff marginalization
- achieve better organization-wide alignment
- develop shared leadership
- define and create an HR/IT governance model

The success of IT/HR alignment should follow practices similar to those I instituted with the line managers at Ravell, specifically the following:

- Successful HR/IT integration requires organizational learning techniques.
- Alignment requires an understanding of the relationship between IT investments and business strategy.
- An integration of IT can create new organizational cultures and structures.
- HR/IT alignment will likely continue to be dynamic in nature, and evolve at an accelerated pace.

The oversight of not integrating better with HR cost IT dearly at Ravell. HR became an undisclosed enemy—that is, a negative force against the entire integration. I discovered this problem only later, and was never able to bring the HR department into the fold. Without HR being part of the learning organization, IT staff continued to struggle with aligning their professional positions with those of the other departments. Fortunately, within two years the HR vice president retired, which inevitably opened the doors for a new start.

In large IT organizations, it is not unusual to have an HR member assigned to focus specifically on IT needs. Typically, it is a joint position in which the HR individual in essence works for the IT executive. This is an effective alternative in that the HR person becomes versed in IT needs and can properly represent IT in the area of head count needs and specific titles. Furthermore, the unique aspect of IT organizations is in the hybrid nature of their staff. Typically, a number of IT staff members are consultants, a situation that presents problems similar to the one I encountered at Ravell—that is, the resentment of not really being part of the organization. Another issue is that many IT staff members are outsourced across the globe, a situation that brings its own set of challenges. In addition, the role of HR usually involves ensuring compliance with various regulations. For example, in many organizations, a consultant is permitted to work on site for only one year before U.S. government regulations force the company to hire them as employees. The HR function must work closely with IT to enforce these regulations. Yet another important component of IT and HR collaboration is talent management. That is, HR must work closely with IT to understand new roles and responsibilities as they develop in the organization. Another challenge is the integration of technology into the day-to-day business of a company, and the question of where IT talent should be dispersed throughout the organization. Given this complex set of challenges, IT alone cannot facilitate or properly represent itself, unless it aligns with the HR departments. This becomes further complex with the proliferation of IT virtual teams across the globe that create complex structures that often have different HR ramifications, both legally and culturally. Virtual team management is discussed further in the book.

Conclusion

This case study shows that strategic integration of technical resources into core business units can be accomplished, by using those aspects of organizational learning that promote reflection in action. This kind of integration also requires something of a concomitant form of assimilation, on the cultural level (see Chapter 3). Reflective thinking fosters the development of a learning organization, which in turn allows for the integration of the "other" in its various organizational manifestations. The experience of this case study also shows that the success of organizational learning will depend on the degree of cross fertilization achievable in terms of individual values and on the ability of the community to combine new concepts and beliefs, to form a hybrid culture. Such a new culture prospers with the use of organizational learning strategies to enable it to share ideas, accept mistakes, and learn to know less as a regular part their discourse and practice in their day-to-day operations.

Another important conclusion from the Ravell experience is that time is an important factor to the success of organizational learning approaches. One way of dealing with the problem of time is with patience—something that many organizations do not have. Another element of success came in the acceleration of events (such as the relocation at Ravell), which can foster a quicker learning cycle and helps us see results faster. Unfortunately, impatience with using organizational learning methods is not an acceptable approach because it will not render results that change individual and organizational behavior. Indeed, I almost changed my approach when I did not get the results I had hoped for early in the Ravell engagement. Nevertheless, my persistence paid off. Finally, the belief that replacing the staff, as opposed to investing in its knowledge, results from a faulty generalization. I found that most of the IT staff had much to contribute to the organization and, ultimately, to help transform the culture. Subsequent chapters of this book build on the Ravell experience and discuss specific methods for integrating organizational learning and IT in ways that can improve competitive advantage.

Another recent perception, which I discuss further in Chapter 4, is the commitment to "complete" integration. Simply put, IT cannot select which departments to work with, or choose to participate only with line managers; as they say, it is "all or nothing at all." Furthermore, as Friedman (2007, p. 8) states "The world is flat." Certainly, part of the "flattening" of the world has been initiated by technology, but it has also created overwhelming challenges for seamless integration of technology within all operations. The flattening of the world has created yet another opportunity for IT to better integrate itself into what is now an everyday challenge for all organizations.

2

THE IT DILEMMA

Introduction

We have seen much discussion in recent writing about how information technology has become an increasingly significant component of corporate business strategy and organizational structure (Bradley & Nolan, 1998; Levine et al., 2000; Siebel, 1999). But, do we know about the ways in which this significance takes shape? Specifically, what are the perceptions and realities regarding the importance of technology from organization leaders, business managers, and core operations personnel? Furthermore, what forms of participation should IT assume within the rest of the organization?

The isolation of IT professionals within their companies often prevents them from becoming active participants in the organization. Technology personnel have long been criticized for their inability to function as part of the business and are often seen as a group falling outside business cultural norms (Schein, 1992). They are frequently stereotyped as "techies" and segregated into areas of the business where they become marginalized and isolated from the rest of the organization. It is my experience, based on case studies such as the one reviewed in Chapter 1 (the Ravell Corporation), that if an organization wishes to absorb its IT department into its core culture, and if it wishes to do so successfully, the company as a whole must be prepared to consider structural changes and to seriously consider using organizational learning approaches.

The assimilation of technical people into an organization presents a special challenge in the development of true organizational learning practices (developed more fully in Chapter 3). This challenge stems from the historical separation of a special group that is seen as standing outside the everyday concerns of the business. IT is generally acknowledged as having a key support function in the organization as a whole. However, empirical studies have shown that it is a challenging

21

endeavor to successfully integrate IT personnel into the learning fold and to do so in such a way that they not only are accepted, but also understood to be an important part of the social and cultural structure of the business (Allen & Morton, 1994; Cassidy, 1998; Langer, 2007; Schein, 1992; Yourdon, 1998).

In his book *In Over Our Heads,* Kegan (1994) discusses the challenges of dealing with individual difference. IT personnel have been consistently regarded as "different" fixtures; as outsiders who do not quite fit easily into the mainstream organization. Perhaps, because of their technical practices, which may at times seem "foreign," or because of perceived differences in their values, IT personnel can become marginalized; imagined as outside the core social structures of business. As in any social structure, marginalization can result in the withdrawal of the individual from the community (Schlossberg, 1989). As a result, many organizations are choosing to outsource their IT services rather than confront and address the issues of cultural absorption and organizational learning. The outsourcing alternative tends to further distance the IT function from the core organization, thus increasing the effects of marginalization. Not only does the outsourcing of IT personnel separate them further from their peers, but it also invariably robs the organization of a potentially important contributor to the social growth and organizational learning of the business. For example, technology personnel should be able to offer insight into how technology can support further growth and learning within the organization. In addition, IT personnel are usually trained to take a logical approach to problem solving; as a result, they should be able to offer a complementary focus on learning. Hence, the integration of IT staff members into the larger business culture can offer significant benefits to an organization in terms of learning and organizational growth.

Some organizations have attempted to improve communications between IT and non-IT personnel through the use of an intermediary who can communicate easily with both groups. This intermediary is known in many organizations as the *business analyst.* Typically, the business analyst will take responsibility for the interface between IT and the larger business community. Although a business analyst may help facilitate communication between IT and non-IT personnel, this arrangement cannot help but carry the implication that different "languages" are spoken by these two groups and, by extension, that direct communication is not possible. Therefore, the use of such an intermediary suffers the danger of failing to promote integration between IT and the rest of the organization; in fact, it may serve to keep the two camps separate. True integration, in the form of direct contact between IT and non-IT personnel, represents a greater challenge for an organization than this remedy would suggest.

Recent Background

Since the 1990s, IT has been seen as a kind of variable that possesses the great potential to reinvent business. Aspects of this promise affected many of the core business rules used by successful chief executives and business managers. While organizations have used IT for the processing of information, decision-support processing, and order processing, the impact of the Internet and e-commerce systems has initiated revolutionary responses in every business sector. This economic phenomenon became especially self-evident with the formation of dot-coms in the mid- and late 1990s. The advent of this phenomenon stressed the need to challenge fundamental business concepts. Many financial wizards surmised that new technologies were indeed changing the very infrastructure of business, affecting how businesses would operate and compete in the new millennium. Much of this hoopla seemed justified by the extraordinary potential that technology offered, particularly with respect to the revolutionizing of old-line marketing principles, for it was technology that came to violate what was previously thought to be protected market conditions and sectors. Technology came to reinvent these business markets and to allow new competitors to cross market in sectors they otherwise could not have entered.

With this new excitement also came fear—fear that fostered unnatural and accelerated entry into technology because any delay might sacrifice important new market opportunities. Violating some of their traditional principles, many firms invested in creating new organizations that would "incubate" and eventually, capture large market segments using the Internet as the delivery vehicle. By 2000, many of these dot-coms were in trouble, and it became clear that their notion of new business models based on the Internet contained significant flaws and shortfalls. As a result of this crisis, the role and valuation of IT is again going through a transformation and once more we are skeptical about the value IT can provide a business and about the way to measure the contributions of IT.

IT in the Organizational Context

Technology not only plays a significant role in workplace operations, but also continues to increase its relevance among other traditional components of any business, such as operations, accounting, and marketing (Earl, 1996b; Langer, 2001a; Schein, 1992). Given this increasing relevance, IT gains significance in relation to

1. The impact it bears on organizational structure
2. The role it can assume in business strategy
3. The ways in which it can be evaluated
4. The extent to which chief executives feel the need to manage operational knowledge and thus to manage IT effectively

IT and Organizational Structure

Sampler's (1996) research explores the relationship between IT and organizational structure. His study indicated that there is no clear-cut relationship that has been established between the two. However, he concluded that there are five principal positions that IT can take in this relationship:

1. IT can lead to centralization of organizational control.
2. Conversely, IT can lead to decentralization of organizational control.
3. IT can bear no impact on organizational control, its significance being based on other factors.
4. Organizations and IT can interact in an unpredictable manner.
5. IT can enable new organizational arrangements, such as networked or virtual organizations.

According to Sampler (1996), the pursuit of explanatory models for the relationship between IT and organizational structure continues to be a challenge, especially since IT plays dual roles. On the one hand, it enhances and constrains the capabilities of workers within the organization, and because of this, it also possesses the ability to create a unique cultural component. While both roles are active, their impact on the organization cannot be predicted; instead, they evolve as unique social norms within the organization. Because IT has changed so dramatically over the past decades, it continues to be difficult to compare prior research on the relationship between IT and organizational structure.

Earl (1996a) studied the effects of applying business process reengineering (BPR) to organizations. BPR is a process that organizations undertake to determine how best to use technology, to improve business performance. Earl concludes that BPR is "an unfortunate title: it does not reflect the complex nature of either the distinctive underpinning concept of BPR [i.e., to reevaluate methods and rules of business operations] or the essential practical challenges to make it happen [i.e., the reality of how one goes about doing that]" (p. 54).

In my 2001 study of the Ravell Corporation ("Fixing Bad Habits," Langer, 2001b), I found that BPR efforts require buy-in from business line managers, and that such efforts inevitably require the adaptation by individuals of different cultural norms and practices.

Schein (1992) recognizes that IT culture represents a subculture in collision with many others within an organization. He concludes that if organizations are to be successful in using new technologies in a global context, they must cope with ceaseless flows of information to ensure organizational health and effectiveness. His research indicates that chief executive officers (CEOs) have been reluctant to implement a new system of technology unless their organizations felt comfortable with it and were ready to use it. While many CEOs were aware of cost and efficiency implications in using IT, few were aware of the potential impact on organizational structure that could result from "adopting an IT view of their organizations" (p. 293). Such results suggest that CEOs need to be more active and more cognizant than they have been of potential shifts in organizational structure when adopting IT opportunities.

The Role of IT in Business Strategy

While many chief executives recognize the importance of IT in the day-to-day operations of their business, their experience with attempting to utilize IT as a strategic business tool, has been frustrating. Typical executive complaints about IT, according to Bensaou and Earl (1998), fall into five problem areas:

1. A lack of correspondence between IT investments and business strategy
2. Inadequate payoff from IT investments
3. The perception of too much "technology for technology's sake"
4. Poor relations between IT specialists and users
5. The creation of system designs that fail to incorporate users' preferences and work habits

McFarlan created a strategic grid (as presented in Applegate et al., 2003) designed to assess the impact of IT on operations and strategy. The grid shows that IT has maximum value when it affects both operations and core business objectives. Based on McFarlan's hypothesis, Applegate et al. established five key questions about IT that may be used by executives to guide strategic decision making:

1. Can IT be used to reengineer core value activities, and change the basis of competition?
2. Can IT change the nature of the relationship, and the balance of power, between buyers and sellers?
3. Can IT build or reduce barriers to entry?
4. Can IT increase or decrease switching costs?
5. Can IT add value to existing products and services, or create new ones?

The research and analysis conducted by McFarlan and Applegate, respectively, suggest that when operational strategy and its results are maximized, IT is given its highest valuation as a tool that can transform the organization. It then receives the maximum focus from senior management and board members. However, Applegate et al. (2003) also focus on the risks of using technology. These risks increase when executives have a poor understanding of competitive dynamics, when they fail to understand the long-term implications of a strategic system that they have launched, or when they fail to account for the time, effort, and cost required to ensure user adoption, assimilation, and effective utilization. Applegate's conclusion underscores the need for IT management to educate senior management, so that the latter will understand the appropriate indicators for what can maximize or minimize their investments in technology.

Szulanski and Amin (2000) claim that while emerging technologies shrink the window in which any given strategy can be implemented, if the strategy is well thought out, it can remain viable. Mintzberg's (1987) research suggests that it would be useful to think of strategy as an art, not a science. This perspective is especially true in situations of uncertainty. The rapidly changing pace of emerging technologies, we know, puts a strain on established approaches to strategy—that is to say, it becomes increasingly difficult to find comfortable implementation of technological strategies in such times of fast-moving environments, requiring sophisticated organizational infrastructure and capabilities.

Ways of Evaluating IT

Firms have been challenged to find a way to best evaluate IT, particularly using traditional return on investment (ROI) approaches. Unfortunately, in this regard, many components of IT do not generate direct returns. Cost allocations based on overhead formulas (e.g., costs of IT as a percentage of revenues) are not applicable to most IT spending needs. Lucas (1999) established nonmonetary methods for evaluating IT. His concept of *conversion effectiveness* places value on the ability of IT to complete its projects on time and within its budgets. This alone is a sufficient factor for providing ROI, assuming that the project was approved for valid business reasons. He called this overall process for evaluation the "garbage can" model. It allows organizations to present IT needs through a funneling pipeline of conversion effectiveness that filters out poor technology plans and that can determine which projects will render direct and indirect benefits to the organization. Indirect returns, according to Lucas, are those that do not provide directly measurable monetary returns but do provide significant value that can be measured using his IT investment opportunities matrix. Utilizing statistical probabilities of returns, the opportunities matrix provides an effective tool for evaluating the impact of indirect returns.

Executive Knowledge and Management of IT

While much literature and research have been produced on how IT needs to participate in and bring value to an organization, there has been relatively little analysis conducted on what non-IT chief executives need to know about technology. Applegate et al. (2003) suggest that non-IT executives need to understand how to differentiate new technologies from older ones, and how to gauge the expected impact of these technologies on the businesses, in which the firm competes for market share. This is to say that technology can change the relationship between customer and vendor, and thus, should be examined as a potential for providing competitive advantage. The authors state that non-IT business executives must become more comfortable with technology by actively participating in technology decisions rather than delegating them to others. They need to question experts as they would in the financial areas of their businesses. Lou Gerstner, former CEO of IBM, is a good example of a non-IT chief executive who acquired sufficient knowledge and understanding of a technology firm. He was then able to form a team of executives who better understood how to develop the products, services, and overall business strategy of the firm.

Allen and Percival (2000) also investigate the importance of non-IT executive knowledge and participation with IT: "If the firm lacks the necessary vision, insights, skills, or core competencies, it may be unwise to invest in the hottest [IT] growth market" (p. 295). The authors point out that success in using emerging technologies is different from success in other traditional areas of business. They concluded that non-IT managers need to carefully consider expected synergies to determine whether an IT investment can be realized and, especially, whether it is efficient to earn cost of capital.

Recent studies have focused on four important components in the linking of technology and business: its relationship to organizational structure, its role in business strategy, the means of its evaluation, and the extent of non-IT executive knowledge in technology. The challenge in determining the best organizational structure for IT is posed by the accelerating technological advances since the 1970s and by the difficulty in comparing organizational models to consistent business cases. Consequently, there is no single organizational structure that has been adopted by businesses.

While most chief executives understand the importance of using technology as part of their business strategy, they express frustration in determining how to effectively implement a technology-based strategic approach. This frustration results from difficulties in understanding how IT investments relate to other strategic business issues, from difficulty in assessing payoff and performance of IT generally and from perceived poor relations between IT and other departments.

Because most IT projects do not render direct monetary returns, executives find themselves challenged to understand technology investments. They have difficulty measuring value since traditional ROI formulas are not applicable. Thus, executives would do better to focus on valuing technology investments by using methods that can determine payback based on a matrix of indirect returns, which do not always include monetary sources. There is a lack of research on the question of what general knowledge non-IT executives need to have to effectively manage the strategic use of technology within their firms. Non-IT chief executives are often not engaged in day-to-day IT activities, and they often delegate dealing with strategic technology issues to other managers. The remainder of this chapter examines the issues raised by the IT dilemma in its various guises especially as they become relevant to, and are confronted from, the top management or chief executive point of view.

IT: A View from the Top

To investigate further the critical issues facing IT, I conducted a study in which I personally interviewed over 40 chief executives in various industries, including finance/investment, publishing, insurance, wholesale/retail, and hotel management. Executives interviewed were either the CEO or president of their respective corporations. I canvassed a population of New York-based midsize corporations for this interview study. Midsize firms, in our case, comprise businesses of between 200 and 500 employees. Face-to-face interviews were conducted, to allow participants the opportunity to articulate their responses, in contrast to answering printed survey questions; executives were therefore allowed to expand, and clarify, their responses to questions. An interview guide (see questions in Tables 2.1 through 2.3) was designed to raise issues relevant to the challenges of using technology, as reported in the recent research literature, and to

Table 2.1   Perception and Role of IT

QUESTION	ANALYSIS
1. How do you define the role and the mission of IT in your firm?	Fifty-seven percent responded that their IT organizations were reactive and did not really have a mission. Twenty-eight percent had an IT mission that was market driven; that is, their IT departments were responsible for actively participating in marketing and strategic processes.
2. What impact has the Internet had on your business strategy?	Twenty-eight percent felt the impact was insignificant, while 24% felt it was critical. The remaining 48% felt that the impact of the Internet was significant to daily transactions.
3. Does the firm have its own internal software development activity? Do you develop your own in-house software or use software packages?	Seventy-six percent had an internal development organization. Eighty-one percent had internally developed software.
4. What is your opinion of outsourcing? Do you have the need to outsource technology? If so, how is this accomplished?	Sixty-two percent had outsourced certain aspects of their technology needs.
5. Do you use consultants to help formulate the role of IT? If yes, what specific roles do they play? If not, why?	Sixty-two percent of the participants used consultants to assist them in formulating the role of IT.
6. Do you feel that IT will become more important to the strategy of the business? If yes, why?	Eighty-five percent felt that IT had recently become more important to the strategic planning of the business.
7. How is the IT department viewed by other departments? Is the IT department liked, or is it marginalized?	Twenty-nine percent felt that IT was still marginalized. Another 29% felt it was not very integrated. Thirty-eight percent felt IT was sufficiently integrated within the organization, but only one chief executive felt that IT was very integrated with the culture of his firm.
8. Do you feel there is too much "hype" about the importance and role of technology?	Fifty-three percent felt that there was no hype. However, 32% felt that there were levels of hype attributed to the role of technology; 10% felt it was "all hype."
9. Have the role and the uses of technology in the firm significantly changed over the last 5 years? If so, what are the salient changes?	Fourteen percent felt little had changed, whereas 43% stated that there were moderate changes. Thirty-eight percent stated there was significant change.

consider significant phenomena, that could affect changes in the uses of technology, such as the Internet. The interview discussions focused on three sections: (1) chief executive perception of the role of IT, (2) management and strategic issues, and (3) measuring IT performance and activities. The results of the interviews are summarized next.

Table 2.2　Management and Strategic Issues

QUESTION	ANALYSIS
1. What is the most senior title held by someone in IT? Where does this person rank on the organization hierarchy?	Sixty-six percent called the highest position chief information officer (CIO). Ten percent used managing director, while 24% used director as the highest title.
2. Does IT management ultimately report to you?	Fifty percent of IT leaders reported directly to the chief executive (CEO). The other half reported to either the chief financial officer (CFO) or the chief operating officer (COO).
3. How active are you in working with IT issues?	Fifty-seven percent stated that they are very active—on a weekly basis. Thirty-eight percent were less active or inconsistently involved, usually stepping in when an issue becomes problematic.
4. Do you discuss IT strategy with your peers from other firms?	Eighty-one percent did not communicate with peers at all. Only 10% actively engaged in peer-to-peer communication about IT strategy.
5. Do IT issues get raised at board, marketing, and/or strategy meetings?	Eighty-six percent confirmed that IT issues were regularly discussed at board meetings. However, only 57% acknowledged IT discussion during marketing meetings, and only 38% confirmed like discussions at strategic sessions.
6. How critical is IT to the day-to-day business?	Eighty-two percent of the chief executives felt it was very significant or critical to the business.

Table 2.3　Measuring IT Performance and Activities

QUESTION	ANALYSIS
1. Do you have any view of how IT should be measured and accounted for?	Sixty-two percent stated that they had a view on measurement; however, there was significant variation in how executives defined measurement.
2. Are you satisfied with IT performance in the firm?	There was significant variation in IT satisfaction. Only 19% were very satisfied. Thirty-three percent were satisfied, another 33% were less satisfied, and 14% were dissatisfied.
3. How do you budget IT costs? Is it based on a percentage of gross revenues?	Fifty-seven percent stated that they did not use gross revenues in their budgeting methodologies.
4. To what extent do you perceive technology as a means of increasing marketing or productivity or both?	Seventy-one percent felt that technology was a significant means of increasing both marketing and productivity in their firms.
5. Are Internet/Web marketing activities part of the IT function?	Only 24% stated that Internet/Web marketing efforts reported directly to the IT organization.

Section 1: Chief Executive Perception of the Role of IT

This section of the interview focuses on chief executive perceptions of the role of IT within the firm. For the first question, about the role and mission of IT, over half of the interviewees responded in ways that suggested their IT organizations were reactive, without a strategic mission. One executive admitted, "IT is not really defined. I guess its mission is to meet our strategic goals and increase profitability." Another response betrays a narrowly construed understanding of its potential: "The mission is that things must work—zero tolerance for failure." These two responses typify the vague and generalized perception that IT "has no explicit mission" except to advance the important overall mission of the business itself. Little over a quarter of respondents could confirm a market-driven role for IT; that is, actively participating in marketing and strategic processes. Question 2, regarding the impact of the Internet on business strategy, drew mixed responses. Some of these revealed the deeply reflective challenges posed by the Internet: "I feel the Internet forces us to take a longer-term view and a sharper focus to our business." Others emphasized its transformative potential: "The Internet is key to decentralization of our offices and business strategy."

Questions 3 and 4 focused on the extent to which firms have their own software development staffs, whether they use internally developed or packaged software, and whether they outsource IT services. Control over internal development of systems and applications remained important to the majority of chief executives: "I do not like outsourcing—surrender control, and it's hard to bring back." Almost two-thirds of the participants employed consultants to assist them in formulating the role of IT within their firms but not always without reservation: "Whenever we have a significant design issue we bring in consultants to help us—but not to do actual development work." Only a few were downright skeptical: "I try to avoid consultants—what is their motivation?" The perception of outsourcing is still low in midsize firms, as compared to the recent increase in IT outsourcing abroad. The lower use could be related to the initial costs and management overheads that are required to properly implement outsource operations in foreign countries.

A great majority of chief executives recognized some form of the strategic importance of IT to business planning: "More of our business is related to technology and therefore I believe IT is more important to strategic planning." Still, this sense of importance remained somewhat intuitive: "I cannot quantify how IT will become more strategic to the business planning—but I sense that job functions will be dramatically altered." In terms of how IT is viewed by other departments within the firm, responses were varied. A little over a third of respondents felt IT was reasonably integrated within the organization: "The IT department is vitally important—but rarely noticed." The majority of respondents, however, recognized a need for greater integration: "IT was marginalized—but it is changing. While IT drives the system—it needs to drive more of the business." Some articulated clearly the perceived problems: "IT needs to be more proactive—they do not seem to have good interpersonal skills and do not understand corporate politics." A few expressed a sense of misgiving ("IT people are strange—personality is an issue") and even a sense of hopelessness: "People hate IT—particularly over the sensitivity of the data. IT sometimes is viewed as misfits and incompetent."

Question eight asked participants whether they felt there was too much "hype" attributed to the importance of technology in business. Over half responded in the negative, although not without reservation: "I do not think there is too much hype—but I am disappointed. I had hoped that technology at this point would have reduced paper, decreased cost—it just has not happened." Others felt that there is indeed some degree of sensationalism: "I definitely think there is too much hype—everyone wants the latest and greatest." Hype in many cases can be related to a function of evaluation, as in this exclamation: "The hype with IT relates more to when will we actually see the value!" The last question in this section asks whether the uses of technology within the firm had significantly changed over the last five years. A majority agreed that it had: "The role of IT has changed significantly in the last five years—we need to stay up-to-date because we want to carry the image that we are 'on the ball'." Many of these stressed the importance of informational flows: "I find the 'I' [information] part to be more and more important and the 'T' [technology] to be diminishing in importance." Some actively downplayed the significance: "I believe in minimizing the amount of technology we use—people get carried away."

Section 2: Management and Strategic Issues

This section focuses on questions pertaining to executive and management organizational concerns. The first and second questions asked executives about the most senior title held by an IT officer and about the reporting structure for IT. Two-thirds of the participants ranked their top IT officer as a chief information officer (CIO). In terms of organizational hierarchy, half of the IT leaders were at the second tier, reporting directly to the CEO or president, while the other half were at the third tier, reporting either to the chief financial officer (CFO) or to the chief operating officer (COO). As one CEO stated, "Most of my activity with IT is through the COO. We have a monthly meeting, and IT is always on the agenda."

The third question asked executives to consider their level of involvement with IT matters. Over half claimed a highly active relationship, engaging on a weekly basis: "I like to have IT people close and in one-on-one interactions. It is not good to have artificial barriers." For some, levels of involvement may be limited: "I am active with IT issues in the sense of setting goals." A third of participants claimed less activity, usually becoming active when difficulties arose. Question four asked whether executives spoke to their peers at other firms about technology issues. A high majority managed to skip this potential for communication with their peers. Only one in 10 actively pursued this matter of engagement.

Question 5 asked about the extent to which IT issues were discussed at board meetings, marketing meetings, and business strategy sessions. Here, a great majority confirmed that there was regular discussion regarding IT concerns, especially at board meetings. A smaller majority attested to IT discussions during marketing meetings. Over a third reported that IT issues maintained a presence at strategic sessions. The higher incidence at board meetings may still be attributable to the effects of Year 2000 (Y2K) preparations. The final question in this section concerned the level of criticality for IT in the day-to-day operations of the business. A high majority of executives responded affirmatively in this regard: "IT is critical to our survival, and its impact on economies of scale is significant."

Section 3: Measuring IT Performance and Activities

This section is concerned with how chief executives measured IT performance and activities within their firms. The first question of this section asked whether executives had a view about how IT performance should be measured. Almost two-thirds affirmed having some formal or informal way of measuring performance: "We have no formal process of measuring IT other than predefined goals, cost constraints, and deadlines." Their responses demonstrated great variation, sometimes leaning on cynicism: "I measure IT by the number of complaints I get." Many were still grappling with this challenge: "Measuring IT is unqualified at this time. I have learned that hours worked is not the way to measure IT—it needs to be more goal-oriented." Most chief executives expressed some degree of quandary: "We do not feel we know enough about how IT should be measured." Question two asked executives to rate their satisfaction with IT performance. Here, also, there was significant variation. A little more than half expressed some degree of satisfaction: "Since 9/11 IT has gained a lot of credibility because of the support that was needed during a difficult time." Slightly fewer than half revealed a degree of dissatisfaction: "We had to overhaul our IT department to make it more customer-service oriented."

Question three concerned budgeting; that is, whether or not chief executives budgeted IT costs as a percentage of gross revenues. Over half denied using gross revenues in their budgeting method: "When handling IT projects we look at it on a request-by-request basis."

The last two questions asked chief executives to assess the impact of technology on marketing and productivity. Almost three quarters of the participants felt that technology represented a significant means of enhancing both marketing and productivity. Some maintained a certainty of objective: "We try to get IT closer to the customer—having them understand the business better." Still, many had a less-defined sense of direction: "I have a fear of being left behind, so I do think IT will become more important to the business." And others remained caught in uncertainty: "I do not fully understand how to use technology in marketing—but I believe it's there." Chief executive certainty, in this matter, also found expression in the opposite direction: "IT will become less important—it will be assumed as a capability and a service that companies provide to their customers." Of the Internet/

Web marketing initiatives, only one quarter of these reported directly to the IT organization: "IT does not drive the Web activities because they do not understand the business." Often, these two were seen as separate or competing entities of technology: "Having Web development report to IT would hinder the Internet business's growth potential." Yet, some might be willing to explore a synergistic potential: "We are still in the early stages of understanding how the Internet relates to our business strategy and how it will affect our product line."

General Results

Section 1 revealed that the matter of defining a mission for the IT organization remains as unresolved as finding a way to reckon with the potential impact of IT on business strategy. Executives still seemed to be at a loss on the question of how to integrate IT into the workplace—a human resource as well as a strategic issue. There was uncertainty regarding the dependability of the technology information received. Most agreed, however, in their need for software development departments to support their internally developed software, in their need to outsource certain parts of technology, and in their use of outside consultants to help them formulate the future activities of their IT departments.

Section 2 showed that while the amount of time that executives spent on IT issues varied, there was a positive correlation between a structure in which IT managers reported directly to the chief executive and the degree of activity that executives stated they had with IT matters. Section 3 showed that chief executives understood the potential value that technology can bring to the marketing and productivity of their firms. They did not believe, however, that technology can go unmeasured; there needs to be some rationale for allotting a spending figure in the budget. For most of the firms in this study, the use of the Internet as a technological vehicle for future business was not determined by IT. This suggests that IT does not manage the marketing aspects of technology, and that it has not achieved significant integration in strategic planning.

Defining the IT Dilemma

The variations found in this study in terms of where IT reports, how it is measured, and how its mission is defined were consistent with existing research. But, the wide-ranging inconsistencies and uncertainties among executives described here left many of them wondering whether they should be using IT as part of their business strategy and operations. While this quandary does not in itself suggest an inadequacy, it does point to an absence of a "best practices" guideline for using technology strategically. Hence, most businesses lacked a clear plan on how to evolve IT contributions toward business development. Although a majority of respondents felt that IT was critical to the survival of their businesses, the degree of IT assimilation within the core culture of organizations still varied. This suggests that the effects of cultural assimilation lag behind the actual involvement of IT in the strategic direction of the company.

While Sampler (1996) attributes many operational inconsistencies to the changing landscape of technology, the findings of this study suggest that there is also a lack in professional procedures, rules, and established governance, that could support the creation of best practices for the profession. Bensaou and Earl (1998), on the one hand, have addressed this concern by taking a pro-Japanese perspective in extrapolating from five "Western" problems five "general" principles, presumably not culture bound, and thence a set of "best principles" for managing IT. But, Earl et al. (1995), on the other hand, have sidestepped any attempt to incorporate Earl's own inductive approach discussed here; instead, they favor a market management approach, based on a supply-and-demand model to "balance" IT management. Of course, best practices already embody the implicit notion of best principles; however, the problems confronting executives—the need for practical guidelines—remain. For instance, this study shows that IT performance is measured in many different ways. It is this type of practical inconsistency that leaves chief executives with the difficult challenge of understanding how technology decisions can be managed.

On a follow-up call related to this study, for example, a CEO informed me of a practical yet significant difference she had instituted since our interview. She stated:

> The change in reporting has allowed IT to become part of the mainstream vision of the business. It now is a fundamental component of all discussions with human resources, sales and marketing, and accounting. The change in reporting has allowed for the creation of a critical system,

which has generated significant direct revenues for the business. I attribute this to my decision to move the reporting of technology directly to me and to my active participation in the uses of technology in our business.

This is an example of an executive whom Schein (1994) would call a "change agent"—someone who employs "cognitive redefinition through scanning," in this case to elicit the strategic potential of IT. We might also call this activity reflective thinking (Langer, 2001b). Schein's change agents, however, go on to "acknowledge that future generations of CEOs will have been educated much more thoroughly in the possibilities of the computer and IT, thus enabling them to take a hands-on adopter stance" (p. 343). This insight implies a distancing ("future") of present learning responsibilities among current chief executives. The nearer future of this insight may instead be seen in the development of organizational learning.* These are two areas of contemporary research that begin to offer useful models in the pursuit of a best practices approach to the understanding and managing of IT.

If the focus of this latter study was geared toward the evaluation of IT based on the view of the chief executive, it was, indeed, because their views necessarily shape the very direction for the organizations that they manage. Subsequent chapters of this book examine how the various dilemmas surrounding IT that I have discussed here are affecting organizations and how organizational learning practices can help answer many of the issues of today as raised by executives, managers, and operations personnel.

Recent Developments in Operational Excellence

The decline in financial markets in 2009, and the continued increase in mergers and acquisitions due to global competition have created an interesting opportunity for IT that reinforces the need for integration via organizational learning. During difficult economic periods, IT has traditionally been viewed as a cost center and had its operations

---

* My case study "Fixing Bad Habits" (Langer, 2001b) has shown that integrating the practices of reflective thinking, to support the development of organizational learning, has greatly enhanced the adaptation of new technologies, their strategic valuation to the firm, and their assimilation into the social norms of the business.

reduced (I discuss this further in Chapter 3, in which I introduce the concept of drivers and supporters). However, with the growth in the role of technology, IT management has now been asked to help improve efficiency through the use of technology across departments. That is, IT is emerging as an agent for business transformation in a much stronger capacity than ever before. This phenomenon has placed tremendous pressure on the technology executive to align with his or her fellow executives in other departments and to get them to participate in cost reductions by implementing more technology. Naturally, using technology to facilitate cuts to the workforce is often unpopular, and there has been much bitter fallout from such cross-department reductions. Technology executives thus face the challenge of positioning themselves as the agents of a necessary change. However, operational excellence is broader than just cutting costs and changing the way things operate; it is about doing things efficiently and with quality measures across corporate operations. Now that technology affects every aspect of operations, it makes sense to charge technology executives with a major responsibility to get it accomplished.

The assimilation of technology as a core part of the entire organization is now paramount for survival, and the technology executive of today and certainly tomorrow will be one who understands that operational excellence through efficiency must be accomplished by educating business units in self-managing the process. The IT executive, then, supports the activity as a leader, not as a cost cutter who invades the business. The two approaches are very different, and adopting the former can result in significant long-term results in strategic alignment.

My interviews with CEOs supported this notion: The CEO does not want to be the negotiator; change must be evolutionary within the business units themselves. While taking this kind of role in organizational change presents a new dilemma for IT, it can also be an opportunity for IT to position itself successfully within the organization.

3

TECHNOLOGY AS A VARIABLE AND RESPONSIVE ORGANIZATIONAL DYNAMISM

Introduction

This chapter focuses on defining the components of technology and how they affect corporate organizations. In other words, if we step back momentarily from the specific challenges that information technology (IT) poses, we might ask the following: What are the generic aspects of technology that have made it an integral part of strategic and competitive advantage for many organizations? How do organizations respond to these generic aspects as catalysts of change? Furthermore, how do we objectively view the role of technology in this context, and how should organizations adjust to its short- and long-term impacts?

Technological Dynamism

To begin, technology can be regarded as a variable, independent of others, that contributes to the life of a business operation. It is capable of producing an overall, totalizing, yet distinctive, effect on organizations—it has the unique capacity to create accelerations of corporate events in an unpredictable way. Technology, in its aspect of unpredictability, is necessarily a variable, and in its capacity as accelerator—its tendency to produce change or advance—it is dynamic. My contention is that, as a dynamic kind of variable, technology, via responsive handling or management, can be tapped to play a special role in organizational development. It can be pressed into service as the dynamic catalyst that helps bring organizations to maturity in dealing not only with new technological quandaries, but also with other agents of change. Change generates new knowledge, which in turn requires a structure of learning that should, if managed properly, result in transformative behavior, supporting the continued evolution of organizational culture. Specifically, technology speeds up events, such as the expectation of getting a response to an e-mail, and requires organizations to respond to them in ever-quickening time frames. Such events are not as predictable as those experienced by individuals in organizations prior to the advent of new technologies—particularly with the meteoric advance of the Internet. In viewing technology then as a dynamic variable, and one that requires systemic and cultural organizational change, we may regard it as an inherent, internal driving force—a form of technological dynamism.

Dynamism is defined as a process or mechanism responsible for the development or motion of a system. *Technological dynamism* characterizes the unpredictable and accelerated ways in which technology, specifically, can change strategic planning and organizational behavior/culture. This change is based on the acceleration of events and interactions within organizations, which in turn create the need to better empower individuals and departments. Another way of understanding technological dynamism is to think of it as an internal drive recognized by the symptoms it produces. The new events and interactions brought about by technology are symptoms of the dynamism that technology manifests. The next section discusses how organizations can begin to make this inherent dynamism work in their favor on different levels.

Responsive Organizational Dynamism

The technological dynamism at work in organizations has the power to disrupt any antecedent sense of comfortable equilibrium or an unwelcome sense of stasis. It also upsets the balance among the various factors and relationships that pertain to the question of how we might integrate new technologies into the business—a question of what we will call *strategic integration*—and how we assimilate the cultural changes they bring about organizationally—a question of what we call *cultural assimilation*. Managing the dynamism, therefore, is a way of managing the effects of technology. I propose that these organizational ripples, these precipitous events and interactions, can be addressed in specific ways at the organizational management level. The set of integrative responses to the challenges raised by technology is what I am calling *responsive organizational dynamism*, which will also receive further explication in the next few chapters. For now, we need to elaborate the two distinct categories that present themselves in response to technological dynamism: strategic integration and cultural assimilation. Figure 3.1 diagrams the relationships.

Strategic Integration

Strategic integration is a process that addresses the business-strategic impact of technology on organizational processes. That is, the business-strategic impact of technology requires immediate organizational responses and in some instances zero latency. Strategic integration recognizes the need to scale resources across traditional business–geographic boundaries, to redefine the value chain in the life cycle of a product or service line, and generally to foster more agile business processes (Murphy, 2002). Strategic integration, then,

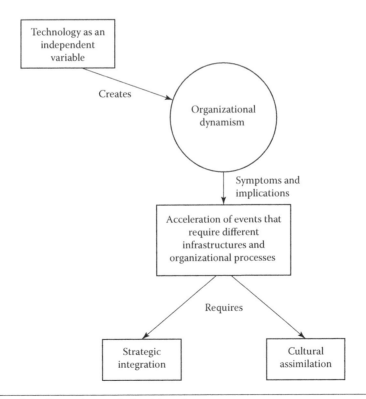

Figure 3.1 Responsive organizational dynamism.

is a way to address the changing requirements of business processes caused by the sharp increases in uses of technology. Evolving technologies have become catalysts for competitive initiatives that create new and different ways to determine successful business investment. Thus, there is a dynamic business variable that drives the need for technology infrastructures capable of greater flexibility and of exhibiting greater integration with all business operations.

Historically, organizational experiences with IT investment have resulted in two phases of measured returns. The first phase often shows negative or declining productivity as a result of the investment; in the second phase, we often see a lagging of, although eventual return to, productivity. The lack of returns in the first phase has been attributed to the nature of the early stages of technology exploration and experimentation, which tend to slow the process of organizational adaptation to technology. The production phase then lags behind the ability of the organization to integrate new technologies with its existing processes. Another complication posed by technological dynamism via the process of strategic integration is a phenomenon we can call *factors of multiplicity*—essentially, what happens when several new technology opportunities overlap and create myriad projects that are in various phases of their developmental life cycle. Furthermore, the problem is compounded by lagging returns in productivity, which are complicated to track and to represent to management. Thus, it is important that organizations find ways to shorten the period between investment and technology's effective deployment. Murphy (2002) identifies several factors that are critical to bridging this delta:

1. Identifying the processes that can provide acceptable business returns from new technological investments
2. Establishing methodologies that can determine these processes
3. Finding ways to actually perform and realize expected benefits
4. Integrating IT projects with other projects
5. Adjusting project objectives when changes in the business require them

Technology complicates these actions, making them more difficult to resolve; hence the need to manage the complications. To tackle these compounded concerns, strategic integration can shorten life cycle maturation by focusing on the following integrating factors:

- Addressing the weaknesses in management organizations in terms of how to deal with new technologies, and how to better realize business benefits
- Providing a mechanism that both enables organizations to deal with accelerated change caused by technological innovations and integrates them into a new cycle of processing and handling change
- Providing a strategic learning framework by which every new technology variable adds to organizational knowledge, particularly using reflective practices (see Chapter 4)
- Establishing an integrated approach that ties technology accountability to other measurable outcomes using organizational learning techniques and theories

To realize these objectives, organizations must be able to

- Create dynamic internal processes that can function on a daily basis to deal with understanding the potential fit of new technologies and their overall value to the business
- Provide the discourse to bridge the gaps between IT- and non-IT-related investments and uses into an integrated system
- Monitor investments and determine modifications to the life cycle
- Implement various organizational learning practices, including learning organization, knowledge management, change management, and communities of practice, all of which help foster strategic thinking and learning that can be linked to performance (Gephardt & Marsick, 2003)

Another important aspect of strategic integration is what Murphy (2002) calls "consequential interoperability," in which "the consequences of a business process" are understood to "dynamically trigger integration" (p. 31). This integration occurs in what he calls the five pillars of benefits realization:

1. *Strategic alignment*: The alignment of IT strategically with business goals and objectives.
2. *Business process impact*: The impact on the need for the organization to redesign business processes and integrate them with new technologies.

3. *Architecture*: The actual technological integration of applications, databases, and networks to facilitate and support implementation.
4. *Payback*: The basis for computing return on investment (ROI) from both direct and indirect perspectives.
5. *Risk*: Identifying the exposure for underachievement or failure in the technology investment.

Murphy's (2002) pillars are useful in helping us understand how technology can engender the need for responsive organizational dynamism (ROD), especially as it bears on issues of strategic integration. They also help us understand what becomes the strategic integration component of ROD. His theory on strategic alignment and business process impact supports the notion that IT will increasingly serve as an undergirding force, one that will drive enterprise growth by identifying the initiators (such as e-business on the Internet) that best fit business goals. Many of these initiators will be accelerated by the growing use of e-business, which becomes the very driver of many new market realignments. This e-business realignment will require the ongoing involvement of executives, business managers, and IT managers. In fact, the Gartner Group forecasted that 70% of new software application investments and 5% of new infrastructure expenditures by 2005 would be driven by e-business. Indeed, this has occurred and continues to expand.

The combination of evolving business drivers with accelerated and changing customer demands has created a business revolution that best defines the imperative of the strategic integration component of ROD. The changing and accelerated way businesses deal with their customers and vendors requires a new strategic integration to become a reality rather than remain a concept discussed but affecting little action. Without action directed toward new strategic integration, organizations would lose competitive advantage, which would affect profits. Most experts see e-business as the mechanism that will ultimately require the integrated business processes to be realigned, thus providing value to customers and modifying the customer–vendor relationship. The driving force behind this realignment emanates from the Internet, which serves as the principle accelerator of the change in transactions across all businesses. The general need to optimize resources forces organizations to rethink and to realign business processes to gain access to new business markets.

Murphy's (2002) pillar of architecture brings out yet another aspect of ROD. By *architecture* we mean the focus on the effects that technology has on existing computer applications or legacy systems (old existing systems). Technology requires existing IT systems to be modified or replacement systems to be created that will mirror the new business realignments. These changes respond to the forces of strategic integration and require business process reengineering (BPR) activities, which represent the reevaluation of existing systems based on changing business requirements. It is important to keep in mind the acceleration factors of technology and to recognize the amount of organizational effort and time that such projects take to complete. We must ask the following question: How might organizations respond to these continual requirements to modify existing processes? I discuss in other chapters how ROD represents the answer to this question.

Murphy's (2002) pillar of direct return is somewhat limited and narrow because not all IT value can be associated with direct returns, but it is important to discuss. Technology acceleration is forcing organizations to deal with broader issues surrounding what represents a return from an investment. The value of strategic integration relies heavily on the ability of technology to encapsulate itself within other departments where it ultimately provides the value. We show in Chapter 4 that this issue also has significance in organizational formation. What this means is simply that value can be best determined within individual business units at the microlevel and that these appropriate-level business units also need to make the case for why certain investments need to be pursued. There are also paybacks that are indirect; for example, Lucas (1999) demonstrates that many technology investments are nonmonetary. The IT department (among others) becomes susceptible to great scrutiny and subject to budgetary cutbacks during economically difficult times. This does not suggest that IT "hide" itself but rather that its investment be integrated within the unit where it provides the most benefit. Notwithstanding the challenge to map IT expenditures to their related unit, there are always expenses that are central to all departments, such as e-mail and network infrastructure. These types of expenses can rarely provide direct returns and are typically allocated across departments as a cost of doing business.

Because of the increased number of technology opportunities, Murphy's (2002) risk pillar must be a key part of strategic integration. The concept of risk assessment is not new to an organization; however, it is somewhat misunderstood as it relates to technology assessment. Technology assessment, because of the acceleration factor, must be embedded within the strategic decision-making process. This can only be accomplished by having an understanding of how to align technology opportunities for business change and by understanding the cost of forgoing the opportunity as well as the cost of delays in delivery. Many organizations use risk assessment in an unstructured way, which does not provide a consistent framework to dynamically deal with emerging technologies. Furthermore, such assessment needs to be managed at all levels in the organization as opposed to being an event-driven activity controlled only by executives.

Summary

Strategic integration represents the objective of dealing with emerging technologies on a regular basis. It is an outcome of ROD, and it requires organizations to deal with a variable, that forces acceleration of decisions in an unpredictable fashion. Strategic integration would require businesses to realign the ways in which they include technology in strategic decision making.

Cultural Assimilation

Cultural assimilation is a process that focuses on the organizational aspects of how technology is internally organized, including the role of the IT department, and how it is assimilated within the organization as a whole. The inherent, contemporary reality of technological dynamism requires not only strategic but also cultural change. This reality demands that IT organizations connect to all aspects of the business. Such affiliation would foster a more interactive culture rather than one that is regimented and linear, as is too often the case. An interactive culture is one that can respond to emerging technology decisions in an optimally informed way, and one that understands the impact on business performance.

The kind of cultural assimilation elicited by technological dynamism and formalized in ROD is divided into two subcategories: the study of how the IT organization relates and communicates with "others," and the actual displacement or movement of traditional IT staff from an isolated "core" structure to a firm-wide, integrated framework.

IT Organization Communications with "Others"

The Ravell case study shows us the limitations and consequences of an isolated IT department operating within an organization. The case study shows that the isolation of a group can lead to marginalization, which results in the kind of organization in which not all individuals can participate in decision making and implementation, even though such individuals have important knowledge and value. Technological dynamism is forcing IT departments to rethink their strategic position within the organizational structure of their firm. No longer can IT be a stand-alone unit designed just to service outside departments while maintaining its separate identity. The acceleration factors of technology require more dynamic activity within and among departments, which cannot be accomplished through discrete communications between groups. Instead, the need for diverse groups to engage in more integrated discourse, and to share varying levels of technological knowledge, as well as business-end perspectives, requires new organizational structures that will of necessity give birth to a new and evolving business—social culture. Indeed, the need to assimilate technology creates a transformative effect on organizational cultures, the way they are formed and re-formed, and what they will need from IT personnel.

Movement of Traditional IT Staff

To facilitate cultural assimilation from an IT perspective, IT must become better integrated with non-IT personnel. This form of integration can require the actual movement of IT staff into other departments, which begins the process of a true assimilation of resources among business units. While this may seem like the elimination of the integrity or identity of IT, such a loss is far from the case. The elimination of the IT department is not at all what is called for here; on the contrary, the IT department is critical to the function of cultural assimilation. However, the IT department may need to be structured differently from the way it has been so that it can deal primarily with generic infrastructure and support issues, such as e-mail, network architecture, and security. IT personnel who focus on business-specific issues need to become closely aligned with the appropriate units so that ROD can be successfully implemented.

Furthermore, we must acknowledge that, given the wide range of available knowledge about technology, not all technological knowledge emanates from the IT department. The question becomes one of finding the best structure to support a broad assimilation of knowledge about any given technology; then, we should ask how that knowledge can best be utilized by the organization. There is a pitfall in attempting to find a "standard" IT organizational structure that will address the cultural assimilation of technology. Sampler's (1996) research, and my recent research with chief executives, confirms that no such standard structure exists. It is my position that organizations must find their own unique blend, using organizational learning constructs. This simply means that the cultural assimilation of IT may be unique to the organization. What is then more important for the success of organizational development is the process of assimilation as opposed to the transplanting of the structure itself.

Today, many departments still operate within "silos" where they are unable to meet the requirements of the dynamic and unpredictable nature of technology in the business environment. Traditional organizations do not often support the necessary communications needed to implement cultural assimilation across business units. However, business managers can no longer make decisions without considering technology; they will find themselves needing to include IT staff in their decision-making processes. On the other hand, IT departments can no longer make technology-based decisions without concerted efforts toward assimilation (in contrast to occasional partnering or project-driven participation) with other business units. This assimilation becomes mature when new cultures evolve synergistically as opposed to just having multiple cultures that attempt to work in conjunction with each other. The important lesson from Ravell to keep in mind here is that the process of assimilating IT can create new cultures that in turn evolve to better support the requirements established by the dynamism of technology.

Eventually, these new cultural formations will not perceive themselves as functioning within an IT or non-IT decision framework but rather as operating within a more central business operation that understands how to incorporate varying degrees of IT involvement as necessary. Thus, organizational cultures will need to fuse together to respond to new business opportunities and requirements brought about by the ongoing acceleration of technological innovation. This was also best evidenced by subsequent events at Ravell. Three years after the original case study, it became necessary at Ravell to integrate one of its business operations with a particular group of IT staff members. The IT personnel actually transferred to the business unit to maximize the benefits of merging both business and technical cultures. Interestingly, this business unit is currently undergoing cultural assimilation and is developing its own behavioral norms influenced by the new IT staff. However, technology decisions within such groups are not limited to the IT transferred personnel. IT and non-IT staff need to formulate decisions using various organizational learning techniques. These techniques are discussed in the next chapter.

Summary

Without appropriate cultural assimilation, organizations tend to have staff that "take shortcuts, [then] the loudest voice will win the day, ad hoc decisions will be made, accountabilities lost, and lessons from successes and failures will not become part of ... wisdom" (Murphy, 2002, p. 152). As in the case of Ravell Corporation, it is essential, then, to provide for consistent governance that fits the profile of the existing culture or can establish the need for a new culture. While many scholars and managers suggest the need to have a specific entity responsible for IT governance, one that is to be placed within the operating structure of the organization, such an approach creates a fundamental problem. It does not allow staff and managers the opportunity to assimilate technologically driven change and understand how to design a culture that can operate under ROD. In other words, the issue of governance is misinterpreted as a problem of structural positioning or hierarchy when it is really one of cultural assimilation. As a result, many business solutions to technology issues often lean toward the prescriptive, instead of the analytical, in addressing the real problem.

Murphy's (2002) risk pillar theory offers us another important component relevant to cultural assimilation. This approach addresses the concerns that relate to the creation of risk cultures formed to deal with the impact of new systems. New technologies can actually cause changes in cultural assimilation by establishing the need to make certain changes in job descriptions, power structures, career prospects, degree of job security, departmental influence, or ownership of data. Each of these potential risks needs to be factored in as an important part of considering how best to organize and assimilate technology through ROD.

Technology Business Cycle

To better understand technology dynamism, or how technology acts as a dynamic variable, it is necessary to define the specific steps that occur during its evolution in an organization. The evolution or business cycle depicts the sequential steps during the maturation of a new technology from feasibility to implementation and through subsequent evolution. Table 3.1 shows the five components that comprise the cycle: feasibility, measurement, planning, implementation, and evolution.

Table 3.1   Technology Business Cycle

CYCLE COMPONENT	COMPONENT DESCRIPTION
Feasibility	Understanding how to view and evaluate emerging technologies, from a technical and business perspective.
Measurement	Dealing with both the direct monetary returns and indirect nonmonetary returns; establishing driver and support life cycles.
Planning	Understanding how to set up projects, establishing participation across multiple layers of management, including operations and departments.
Implementation	Working with the realities of project management; operating with political factions, constraints; meeting milestones; dealing with setbacks; having the ability to go live with new systems.
Evolution	Understanding how acceptance of new technologies affects cultural change, and how uses of technology will change as individuals and organizations become more knowledgeable about technology, and generate new ideas about how it can be used; objective is established through organizational dynamism, creating new knowledge and an evolving organization.

Feasibility

The stage of feasibility focuses on a number of issues surrounding the practicality of implementing a specific technology. Feasibility addresses the ability to deliver a product when it is needed in comparison to the time it takes to develop it. Risk also plays a role in feasibility assessment; of specific concern is the question of whether it is possible or probable that the product will become obsolete before completion. Cost is certainly a huge factor, but viewed at a "high level" (i.e., at a general cost range), and it is usually geared toward meeting the expected ROI of a firm. The feasibility process must be one that incorporates individuals in a way that allows them to respond to the accelerated and dynamic process brought forth by technological innovations.

Measurement

Measurement is the process of understanding how an investment in technology is calculated, particularly in relation to the ROI of an organization. The complication with technology and measurement is that it is simply not that easy to determine how to calculate such a return. This problem comes up in many of the issues discussed by Lucas (1999) in his book *Information Technology and the Productivity Paradox*. His work addresses many comprehensive issues, surrounding both monetary and nonmonetary ROI, as well as direct versus indirect allocation of IT costs. Aside from these issues, there is the fact that for many investments in technology the attempt to compute ROI may be an inappropriate approach. As stated, Lucas offered a "garbage can" model that advocates trust in the operational management of the business and the formation of IT representatives into productive teams that can assess new technologies as a regular part of business operations. The garbage can is an abstract concept for allowing individuals a place to suggest innovations brought about by technology. The inventory of technology opportunities needs regular evaluation. Lucas does not really offer an explanation of exactly how this process should work internally. ROD, however, provides the strategic processes and organizational–cultural needs that can provide the infrastructure to better understand and

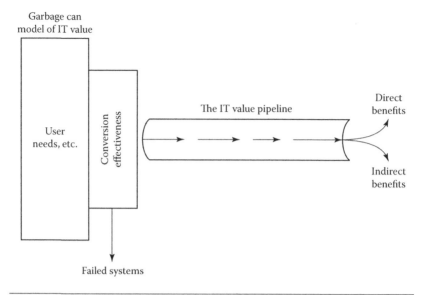

Figure 3.2   Garbage can model of IT value. (From Lucas, H.C., *Information Technology and the Productivity Paradox*. Oxford University Press, New York, 1999.)

evaluate the potential benefits from technological innovations using the garbage can model. The graphic depiction of the model is shown in Figure 3.2.

Planning

Planning requires a defined team of user and IT representatives. This appears to be a simple task, but it is more challenging to understand how such teams should operate, from whom they need support, and what resources they require. Let me be specific. There are a number of varying types of "users" of technology. They typically exist in three tiers: executives, business line managers, and operations users. Each of these individuals offers valuable yet different views of the benefits of technology (Langer, 2002). I define these user tiers as follows:

1. *Executives*: These individuals are often referred to as *executive sponsors*. Their role is twofold. First, they provide input into the system, specifically from the perspective of productivity, ROI, and competitive edge. Second, and perhaps more important, their responsibility is to ensure that users are participating in the requisite manner (i.e., made

to be available, in the right place, etc.). This area can be problematic because internal users are typically busy doing their jobs and sometimes neglect to provide input or to attend project meetings. Furthermore, executive sponsors can help control political agendas that can hurt the success of the project.

2. *Business line managers*: This interface provides the most information from a business unit perspective. These individuals are responsible for two aspects of management. First, they are responsible for the day-to-day productivity of their unit; therefore, they understand the importance of productive teams, and how software can assist in this endeavor. Second, they are responsible for their staff. Thus, line managers need to know how software will affect their operational staff.

3. *Functional users*: These are the individuals in the trenches who understand exactly how processing needs to get done. While their purview of the benefits of the system is relatively narrower than that of the executives and managers, they provide the concrete information that is required to create the feature/functions that make the system usable.

The planning process becomes challenging when attempting to get the three user communities to integrate their needs and "agree to agree" on how a technology project needs to be designed and managed.

Implementation

Implementation is the process of actually using a technology. Implementation of technology systems requires wider integration within the various departments than other systems in an organization because usually multiple business units are affected. Implementation must combine traditional methods of IT processes of development yet integrate them within the constraints, assumptions, and cultural (perhaps political) environments of different departments. Cultural assimilation is therefore required at this stage because it delves into the structure of the internal organization and requires individual participation in every phase of the development and implementation cycle. The following are some of the unique challenges facing the implementation of technological projects:

1. *Project managers as complex managers*: Technology projects require multiple interfaces that often lie outside the traditional user community. They can include interfacing with writers, editors, marketing personnel, customers, and consumers, all of whom are stakeholders in the success of the system.

2. *Shorter and dynamic development schedules*: Due to the dynamic nature of technology, its process of development is less linear than that of others. Because there is less experience in the general user community, and there are more stakeholders, there is a tendency by those in IT, and executives, to underestimate the time and cost to complete the project.

3. *New untested technologies*: There is so much new technology offered to organizations that there is a tendency by IT organizations to implement technologies that have not yet matured—that are not yet the best products they will eventually be.

4. *Degree of scope changes*: Technology, because of its dynamic nature, tends to be prone to *scope creed*—the scope of the original project expanding during development.

5. *Project management*: Project managers need to work closely with internal users, customers, and consumers to advise them on the impact of changes to the project schedule. Unfortunately, scope changes that are influenced by changes in market trends may not be avoidable. Thus, part of a good strategy is to manage scope changes rather than attempt to stop them, which might not be realistic.

6. *Estimating completion time*: IT has always had difficulties in knowing how long it will take to implement a technology. Application systems are even more difficult because of the number of variables and unknowns.

7. *Lack of standards*: The technology industry continues to be a profession that does not have a governing body. Thus, it is impossible to have real enforced standards that other professions enjoy. While there are suggestions for best practices, many of them are unproven and not kept current with

changing developments. Because of the lack of successful application projects, there are few success stories to create new and better sets of best practices.

8. *Less-specialized roles and responsibilities*: The IT team tends to have staff members who have varying responsibilities. Unlike traditional new technology-driven projects, separation of roles and responsibilities is more difficult when operating in more dynamic environments. The reality is that many roles have not been formalized and integrated using something like ROD.

9. *Broad project management responsibilities*: Project management responsibilities need to go beyond those of the traditional IT manager. Project managers are required to provide management services outside the traditional software staff. They need to interact more with internal and external individuals, as well as with non-traditional members of the development team, such as Web text and content staff. Therefore, there are many more obstacles that can cause implementation problems.

Evolution

The many ways to form a technological organization with a natural capacity to evolve have been discussed from an IT perspective in this chapter. However, another important factor is the changing nature of application systems, particularly those that involve e-businesses. E-business systems are those that utilize the Internet and engage in e-commerce activities among vendors, clients, and internal users in the organization. The ways in which e-business systems are built and deployed suggest that they are evolving systems. This means that they have a long life cycle involving ongoing maintenance and enhancement. They are, if you will, "living systems" that evolve in a manner similar to organizational cultures. So, the traditional beginning-to-end life cycle does not apply to an e-business project that must be implemented in inherently ongoing and evolving phases. The important focus is that technology and organizational development have parallel evolutionary processes that need to be in balance with each other. This philosophy is developed further in the next chapter.

Drivers and Supporters

There are essentially two types of generic functions performed by departments in organizations: driver functions and supporter functions. These functions relate to the essential behavior and nature of what a department contributes to the goals of the organization. I first encountered the concept of drivers and supporters at Coopers & Lybrand, which was at that time a Big 8* accounting firm. I studied the formulation of driver versus supporter as it related to the role of our electronic data processing (EDP) department. The firm was attempting to categorize the EDP department as either a driver or a supporter.

Drivers were defined in this instance as those units that engaged in frontline or direct revenue-generating activities. *Supporters* were units that did not generate obvious direct revenues but rather were designed to support frontline activities. For example, operations such as internal accounting, purchasing, or office management were all classified as supporter departments. Supporter departments, due to their nature, were evaluated on their effectiveness and efficiency or economies of scale. In contrast, driver organizations were expected to generate direct revenues and other ROI value for the firm. What was also interesting to me at the time was that drivers were expected to be more daring—since they must inevitably generate returns for the business. As such, drivers engaged in what Bradley and Nolan (1998) coined "sense and respond" behaviors and activities. Let me explain.

Marketing departments often generate new business by investing or "sensing" an opportunity quickly because of competitive forces in the marketplace. Thus, they must sense an opportunity and be allowed to respond to it in a timely fashion. The process of sensing opportunity, and responding with competitive products or services, is a stage in the cycle that organizations need to support. Failures in the cycles of sense and respond are expected. Take, for example, the

---

* The original "Big 8" consisted of the eight large accounting and management consulting firms—Coopers & Lybrand, Arthur Anderson, Touche Ross, Deloitte Haskins & Sells, Arthur Young, Price Waterhouse, Pete Marwick Mitchell, and Ernst and Whinney—until the late 1980s, when these firms began to merge. Today, there are four: Price Waterhouse Coopers, Deloitte & Touche, Ernst & Young, and KPMG (Pete Marwick and others).

launching of new fall television shows. Each of the major stations goes through a process of sensing which shows might be interesting to the viewing audience. They respond, after research and review, with a number of new shows. Inevitably, only a few of these selected shows are actually successful; some fail almost immediately. While relatively few shows succeed, the process is acceptable and is seen by management as the consequence of an appropriate set of steps for competing effectively—even though the percentage of successful new shows is low. Therefore, it is safe to say that driver organizations are expected to engage in high-risk operations, of which many will fail, for the sake of creating ultimately successful products or services.

The preceding example raises two questions: (1) How does sense and respond relate to the world of IT? and (2) Why is it important? IT is unique in that it is both a driver and a supporter. The latter is the generally accepted norm in most firms. Indeed, most IT functions are established to support myriad internal functions, such as

- Accounting and finance
- Data center infrastructure (e-mail, desktop, etc.)
- Enterprise-level application (enterprise resource planning, ERP)
- Customer support (customer relationship management, CRM)
- Web and e-commerce activities

As one would expect, these IT functions are viewed as overhead related, as somewhat of a commodity, and thus are constantly managed on an economy-of-scale basis—that is, how can we make this operation more efficient, with a particular focus on cost containment?

So, what then are IT driver functions? By definition, they are those that engage in direct revenues and identifiable ROI. How do we define such functions in IT because most activities are sheltered under the umbrella of marketing organization domains? (Excluding, of course, software application development firms that engage in marketing for their actual application products.) I define IT driver functions as those projects that, if delivered, would change the relationship between the organization and its customers; that is, those activities that directly affect the classic definition of a market: forces of supply and demand, which are governed by the customer (demand) and the vendor (supplier) relationship. This concept can be shown in the case example that follows.

Santander versus Citibank

Santander Bank, the major bank of Spain, had enjoyed a dominant market share in its home country. Citibank had attempted for years to penetrate Santander's dominance using traditional approaches (opening more branch offices, marketing, etc.) without success, until, that is, they tried online banking. Using technology as a driver, Citibank made significant penetration into the market share of Santander because it changed the customer–vendor relationship. Online banking, in general, has had a significant impact on how the banking industry has established new markets, by changing this relationship. What is also interesting about this case is the way in which Citibank accounted for its investment in online banking; it knows little about its total investment and essentially does not care about its direct payback. Rather, Citibank sees its ROI in a similar way that depicts driver/marketing behavior; the payback is seen in broader terms to affect not only revenue generation, but also customer support and quality recognition.

Information Technology Roles and Responsibilities

The preceding section focuses on how IT can be divided into two distinct kinds of business operations. As such, the roles and responsibilities within IT need to change accordingly and be designed under the auspices of driver and supporter theory. Most traditional IT departments are designed to be supporters, so that they have a close-knit organization that is secure from outside intervention and geared to respond to user needs based on requests. While in many instances this type of formation is acceptable, it is limited in providing the IT department with the proper understanding of the kind of business objectives that require driver-type activities. This was certainly the experience in the Ravell case study. In that instance, I found that making the effort to get IT support personnel "out from their comfortable shells" made a huge difference in providing better service to the organization at large. Because more and more technology is becoming driver essential, this development will require of IT personnel an increasing ability to communicate to managers and executives and to assimilate within other departments.

The Ravell case, however, also brought to light the huge vacuum of IT presence in driver activities. The subsequent chief executive interview study also confirmed that most marketing IT-oriented activities, such as e-business, do not fall under the purview of IT in most organizations. The reasons for this separation are correlated with the lack of IT executive presence within the management team.

Another aspect of driver and supporter functions is the concept of a life cycle. A life cycle, in this respect, refers to the stages that occur before a product or service becomes obsolete. Technology products have a life cycle of value just as any other product or service. It is important not to confuse this life cycle with processes during development as discussed elsewhere in this chapter.

Many technical products are adopted because they are able to deliver value that is typically determined based on ROI calculations. However, as products mature within an organization, they tend to become more of a commodity, and as they are normalized, they tend to become support-oriented. Once they reach the stage of support, the rules of economies of scale become more important and relevant to evaluation. As a product enters the support stage, replacement based on economies of scale can be maximized by outsourcing to an outside vendor who can provide the service cheaper. New technologies then can be expected to follow this kind of life cycle, by which their initial investment requires some level of risk to provide returns to the business. This initial investment is accomplished in ROD using strategic integration. Once the evaluations are completed, driver activities will prevail during the maturation process of the technology, which will also require cultural assimilation. Inevitably, technology will change organizational behavior and structure. However, once the technology is assimilated and organizational behavior and structures are normalized, individuals will use it as a permanent part of their day-to-day operations. Thus, driver activities give way to those of supporters. Senior managers become less involved, and line managers then become the more important group that completes the transition from driver to supporter.

Replacement or Outsource

After the technology is absorbed into operations, executives will seek to maximize the benefit by increased efficiency and effectiveness.

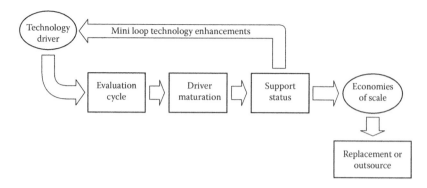

Figure 3.3 Driver-to-supporter life cycle.

Certain product enhancements may be pursued during this phase; they can create "mini-loops" of driver-to-supporter activities. Ultimately, a technology, viewed in terms of its economies of scale and longevity, is considered for replacement or outsourcing. Figure 3.3 graphically shows the cycle.

The final stage of maturity of an evolving driver therefore includes becoming a supporter, at which time it becomes a commodity and, finally, an entity with potential for replacement or outsourcing. The next chapter explores how organizational learning theories can be used to address many of the issues and challenges brought forth in this chapter.

4

ORGANIZATIONAL LEARNING THEORIES AND TECHNOLOGY

Introduction

The purpose of this chapter is to provide readers with an understanding of organizational theory. The chapter covers some aspects of the history and context of organizational learning. It also defines and explains various learning protocols, and how they can be used to promote organizational learning. The overall objective of organizational learning is to support a process that guides individuals, groups, and entire communities through transformation. Indeed, evidence of organizational transformation provides the very proof that learning has occurred, and that changes in behavior are occurring. What is important in this regard is that transformation remains internal to the organization so that it can evolve in a progressive manner while maintaining the valuable knowledge base that is contained within the personnel of an organization. Thus, the purpose of organizational learning is to foster evolutionary transformation that will lead to change in behaviors and that is geared toward improving strategic performance.

Approaches to organizational learning typically address how individuals, groups, and organizations "notice and interpret information and use it to alter their fit with their environments" (Aldrich, 2001, p. 57). As such, however, organizational learning does not direct itself toward, and therefore has not been able to show, an inherent link to success—which is a critical concern for executive management. There are two perspectives on organizational learning theory. On the one hand, the adoptive approach, pioneered by Cyert and March (1963), treats organizations as goal-oriented activity systems. These systems generate learning when repeating experiences that have either succeeded or failed, discarding, of course, processes that have failed.

Knowledge development, on the other hand, treats organizations as sets of interdependent members with shared patterns of cognition and belief (Argyris & Schön, 1996). Knowledge development emphasizes that learning is not limited to simple trial and error, or direct experience. Instead, learning is understood also to be inferential and vicarious; organizations can generate new knowledge through experimentation and creativity. It is the knowledge development perspective that fits conceptually and empirically with work on technological evolution and organizational knowledge creation and deployment (Tushman & Anderson, 1986).

There is a complication in the field of organizational learning over whether it is a technical or social process. Scholars disagree on this point. From the technical perspective, organizational learning is about the effective processing of, interpretation of, and response to information both inside and outside the organization. "An organization is assumed to learn if any of its units acquires knowledge that it recognizes as potentially useful to the organization" (Huber, 1991, p. 89). From the social perspective, on the other hand, comes the concept that learning is "something that takes place not with the heads of individuals, but in the interaction between people" (Easterby-Smith et al., 1999, p. 6). The social approach draws from the notion that patterns of behavior are developed, via patterns of socialization, by evolving tacit knowledge and skills. There is, regrettably, a lack of ongoing empirical investigation in the area of organizational learning pertaining, for example, to in-depth case studies, to micropractices within organizational settings, and to processes that lead to outcomes. Indeed, measuring learning is a difficult process, which is why there is a lack of research that focuses on outputs. As Prange (1999, p. 24) notes: "The multitude of ways in which organizational learning has been classified and used purports an 'organizational learning jungle,' which is becoming progressively dense and impenetrable." Mackenzie (1994, p. 251) laments that what the "scientific community devoted to organizational learning has not produced discernable intellectual progress."

Ultimately, organizational learning must provide transformation that links to performance. Most organizations seeking improved performance expect changes that will support new outcomes. The study of organizational learning needs an overarching framework under which an inquiry into the pivotal issues surrounding organizational change can be organized. Frameworks that support organizational learning, whether their orientation is on individuals, groups, or infrastructure, need to allow for natural evolution within acceptable time frames for the organization. This is the problem of organizational learning theory. It lacks a method of producing measurable results that executives can link to performance. While scholars seek outcomes through strategic learning, there must be tangible evidence of individual and organizational performance to ensure future investments in the concepts of learning. Technology, we should remember, represents the opportunity to provide outcomes through strategic learning that addresses transitions and transformations over a specific life cycle.

We saw this opportunity occur in the Ravell case study; the information technology (IT) department used organizational learning. Specifically, individual reflective practices were used to provide measurable outcomes for the organization. In this case, the outcomes related to a specific event, the physical move of the business to a different location. Another lesson we can derive (with hindsight) from the Ravell experience is that learning was converted to strategic benefit for the organization. The concept of converting learning to strategic benefit was pioneered by Pietersen (2002). He established a strategic learning cycle composed of four component processes that he identified with the action verbs *learn, focus, align,* and *execute.* These are stages in the learning cycle, as follows:

1. *Learn*: Conduct a situation analysis to generate insights into the competitive environment and into the realities of the company.
2. *Focus*: Translate insights into a winning proposition that outlines key priorities for success.
3. *Align*: Align the organization and energize the people behind the new strategic focus.
4. *Execute*: Implement strategy and experiment with new concepts. Interpret results and continue the cycle.

At Ravell, technology assisted in driving the learning cycle because, by its dynamic nature, it mandated the acceleration of the cycle that Pietersen (2002) describes in his stage strategy of implementation. Thus, Ravell required the process Pietersen outlined to occur within

6 months, and therein established the opportunity to provide outcomes. It also altered the culture of the organization (i.e., the evolution in culture was tangible because the transformation was concrete).

We see from the Ravell case that technology represents the best opportunity to apply organizational learning techniques because the use of it requires forms of evolutionary-related change. Organizations are continually seeking to improve their operations and competitive advantage through efficiency and effective processes. As I have discussed in previous chapters, today's businesses are experiencing technological dynamism (defined as causing accelerated and dynamic transformations), and this is due to the advent of technologically driven processes. That is, organizations are experiencing more pressure to change and compete as a result of the accelerations that technology has brought about. Things happen quicker, and more unpredictably, than before. This situation requires organizations to sense the need for change and execute that change. The solution I propose is to tie organizational theory to technological implementation. Another way of defining this issue is to provide an overarching framework that organizes an inquiry into the issues surrounding organizational change.

Another dimension of organizational learning is political. Argyris (1993) and Senge (1990) argue that politics gets "in the way of good learning." In my view, however, the political dimension is very much part of learning. It seems naïve to assume that politics can be eliminated from the daily commerce of organizational communication. Instead, it needs to be incorporated as a factor in organizational learning theory rather than attempting to disavow or eliminate it, which is not realistic. Ravell also revealed that political factors are simply part of the learning process. Recall that during my initial efforts to create a learning organization there were IT staff members who deliberately refused to cooperate, assuming that they could "outlast" me in my interim tenure as IT director. But politics, of course, is not limited to internal department negotiations; it was also a factor at Ravell with, and among, departments outside IT. These interdepartmental relationships applied especially to line managers, who became essential advocates for establishing and sustaining necessary forms of learning at the organizational level. But, not all line managers responded with the same enthusiasm, and a number of them did not display a sense of authentically caring about facilitating synergies across departments.

The irrepressible existence of politics in social organizations, however, must not in itself deter us from implementing organizational learning practices; it simply means that that we must factor it in as part of the equation. At Ravell, I had to work within the constraints of both internal and external politics. Nevertheless, in the end I was able to accomplish the creation of a learning organization. Another way one might look at the road bumps of politics is to assume that they will temporarily delay or slow the implementation of organizational learning initiatives. But, let us make no mistake about the potentially disruptive nature of politics because, as we know, in its extreme cases of inflexibility, it can be damaging.

I have always equated politics with the dilemma of blood cholesterol. We know that there are two types of cholesterol: "good" cholesterol and "bad" cholesterol. We all know that bad cholesterol in your blood can cause heart disease, among other life-threatening conditions. However, good cholesterol is essential to the body. My point is simple; the general word *politics* can have damaging perceptions. When most people discuss the topic of cholesterol, they focus on the bad type, not the good. Such is the same with politics—that is, most individuals discuss the bad type, which often corresponds with their personal experiences. My colleague Professor Lyle Yorks, at Columbia University, often lectures on the importance of politics and its positive aspects for establishing *strategic advocacy*, defined as the ability to establish personal and functional influence through cultivating alliances through defining opportunities for the adding value to either the top or bottom line (Langer & Yorks, 2013). Thus, politics can add value for individuals by allowing them to initiate and influence relationships and conversations with other leaders. This, then, is "good" politics!

North American cultural norms account for much of what goes into organizational learning theory, such as individualism, an emphasis on rationality, and the importance of explicit, empirical information. IT, on the other hand, has a broadening, globalizing effect on organizational learning because of the sheer increase in the number of multicultural organizations created through the expansion of global firms. Thus, technology also affects the social aspects of organizational learning, particularly as it relates to the cultural evolution of communities. Furthermore, technology has shown us that what works in one culture may not work in another. Dana Deasy, the former CIO of the

Americas region/sector for Siemens AG, experienced the difficulties and challenges of introducing technology standards on a global scale. He quickly learned that what worked in North America did not operate with the same expectations in Asia or South America. I discuss Siemens AG as a case study in Chapter 8.

It is my contention, however, that technology can be used as an intervention that can actually increase organizational learning. In effect, the implementation of organizational learning has lacked and has needed concrete systemic processes that show results. A solution to this need can be found, as I have found it, in the incorporation of IT itself into the process of true organizational learning. The problem with IT is that we keep trying to simplify it—trying to reduce its complexity. However, dealing with the what, when, and how of working with technology is complex. Organizations need a kind of mechanism that can provide a way to absorb and learn all of the complex pieces of technology.

It is my position that organizational change often follows learning, which to some extent should be expected. What controls whether change is radical or evolutionary depends on the basis on which new processes are created (Argyris & Schön, 1996; Senge, 1990; Swieringa & Wierdsma, 1992). Indeed, at Ravell the learning followed the Argyris and Schön approach: that radical change occurs when there are major events that support the need for accelerated change. In other words, critical events become catalysts that promote change, through reflection. On the other hand, there can be non-event-related learning, that is not so much radical in nature, as it is evolutionary. Thus, evolutionary learning is characterized as an ongoing process that slowly establishes the need for change over time. This evolutionary learning process compares to what Senge (1990, p. 15) describes as "learning in wholes as opposed to pieces."

This concept of learning is different from an event-driven perspective, and it supports the natural tendency that groups and organizations have to protect themselves from open confrontation and critique. However, technology provides an interesting variable in this regard. It is generally accepted as an agent of change that must be addressed by the organization. I believe that this agency can be seized as an opportunity to promote such change because it establishes a reason why organizations need to deal with the inevitable transitions brought about by technology. Furthermore, as Huysman (1999) points out, the history of organizational learning has not often created measurable improvement, particularly because implementing the theories has not always been efficient or effective. Much of the impetus for implementing a new technology, however, is based on the premise that its use will result in such benefits. Therefore, technology provides compelling reasons for why organizational learning is important: to understand how to deal with agents of change, and to provide ongoing changes in the processes that improve competitive advantage.

There is another intrinsic issue here. Uses of technology have not always resulted in efficient and effective outcomes, particularly as they relate to a firm's expected ROI. In fact, IT projects often cost more than expected and tend to be delivered late. Indeed, research performed by the Gartner Group and *CIO Magazine* (Koch, 1999) reports that 54% of IT projects are late and that 22% are never completed. In May 2009, McGraw reported similar trends, so industry performance has not materially improved. This is certainly a disturbing statistic for a dynamic variable of change that promises outcomes of improved efficiency and effectiveness. The question then is why is this occurring? Many scholars might consider the answer to this question as complex. It is my claim, however, based on my own research, that the lack of organizational learning, both within IT and within other departments, poses, perhaps, the most significant barrier to the success of these projects in terms of timeliness and completion. Langer (2001b) suggests that the inability of IT organizations to understand how to deal with larger communities within the organization and to establish realistic and measurable outcomes are relevant both to many of the core values of organizational learning and to its importance in attaining results. What better opportunity is there to combine the strengths and weaknesses of each of IT and organizational learning?

Perhaps what is most interesting—and, in many ways, lacking within the literature on organizational learning—is the actual way individuals learn. To address organizational learning, I believe it is imperative to address the learning styles of individuals within the organization. One fundamental consideration to take into account is that of individual turnover within departments. Thus, methods to measure or understand organizational learning must incorporate the individual; how the individual learns, and what occurs when individuals change positions or leave, as opposed to solely focusing on the event-driven aspect of evolutionary learning. There are two sociological positions about how individual learning occurs. The first suggests that individual action derives from determining influences in the social system, and the other suggests that it emanates from individual action. The former proposition supports the concept that learning occurs at the organizational, or group level, and the latter supports it at the individual level of action and experience. The "system" argument focuses on learning within the organization as a whole and claims that individual action functions within its boundaries. The "individual" argument claims that learning emanates from the individual first and affects the system as a result of outcomes from individual actions. Determining a balance between individual and organizational learning is an issue debated by scholars and an important one that this book must address.

Why is this issue relevant to the topic of IT and organizational learning? Simply put, understanding the nature of evolving technologies requires that learning—and subsequent learning outcomes—will be heavily affected by the processes in which it is delivered. Therefore, without understanding the dynamics of how individuals and organizations learn, new technologies may be difficult to assimilate because of a lack of process that can determine how they can be best used in the business. What is most important to recognize is the way in which responsive organizational dynamism (ROD) needs both the system and individual approaches. Huysman (1999) suggests (and I agree) that organizational versus individual belief systems are not mutually exclusive pairs but dualities. In this way, organizational processes are not seen as just top-down or bottom-up affairs, but as accumulations of history, assimilated in organizational memory, which structures and positions the agency or capacity for learning. In a similar way, organizational learning can be seen as occurring through the actions of individuals, even when they are constrained by institutional forces. The strategic integration component of ROD lends itself to the system model of learning to the extent that it almost mandates change—change that, if not addressed, will inevitably affect the competitive advantage of the organization. On the other hand, the cultural assimilation component of ROD is also involved because of its effect on individual behavior. Thus, the ROD model needs to be expanded to show the relationship between individual and organizational learning as shown in Figure 4.1.

An essential challenge to technology comes from the fact that organizations are not sure about how to handle its overall potential. Thus, in a paradoxical way, this quandary provides a springboard to learning by utilizing organizational learning theories and concepts to create new knowledge, by learning from experience, and ultimately by linking technology to learning and performance. This perspective can be promoted from within the organization because chief executives are generally open to investing in learning as long as core business principles are not violated. This position is supported by my research with chief executives that I discussed in Chapter 2.

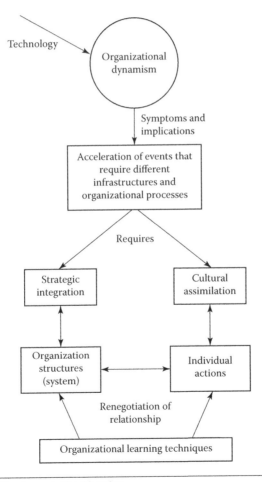

Figure 4.1   ROD and organizational learning.

Organizational learning can also assist in the adoption of technologies by providing a mechanism to help individuals manage change. This notion is consistent with Aldrich (2001), who observes that many organizations reject technology-driven changes or "pioneering ventures," which he called competence-destroying ventures because they threaten existing norms and processes. Organizations would do well to understand the value of technology, particularly for those who adopt it early (early adopters), and how it can lead to competitive advantages. Thus, organizations that position themselves to evolve, to learn, and to create new knowledge are better prepared to foster the handling, absorption, and acceptance of technology-driven change than those that are not. Another way to view this ethic is to recognize that organizations need to be "ready" to deal with change—change that is accelerated by technology innovations. Although Aldrich (2001) notes that organizational learning has not been tied to performance and success, I believe it will be the technology revolution that establishes the catalyst that can tie organizational learning to performance.

The following sections of this chapter expand on the core concept that the success of ROD is dependent on the uses of organizational learning techniques. In each section, I correlate this concept to many of the organizational learning theories and show how they can be tailored and used to provide important outcomes that assist the promotion of both technological innovation and organizational learning.

Learning Organizations

Business strategists have realized that the ability of an organization to learn faster, or "better," than its competitors may indeed be the key to long-term business success (Collis, 1994; Dodgson, 1993; Grant, 1996; Jones, 1975). A *learning organization* is defined as a form of organization that enables, in an active sense, the learning of its members in such a way that it creates positive outcomes, such as innovation, efficiency, improved alignment with the environment, and competitive advantage. As such, a learning organization is one that acquires knowledge from within. Its evolution, then, is primarily driven by itself without the need for interference from outside forces. In this sense, it is a self-perpetuating and self-evolving system of individual and organizational transformations integrated into the daily processes of the organization. It should be, in effect, a part of normal organizational behavior. The focus of organizational learning is not so much on the process of learning but more on the conditions that allow successful outcomes to flourish. Learning organization literature draws from organizational learning theory, particularly as it relates to interventions based on outcomes. This provides an alternative to social approaches.

In reviewing these descriptions of what a learning organization does, and why it is important, we can begin to see that technology may be one of the few agents that can actually show what learning organizations purport to do. Indeed, Ravell created an evolving population that became capable of dealing with environmental changes brought on by technological innovation. The adaptation of these changes created those positive outcomes and improved efficiencies. Without organizational learning, specifically the creation of a learning organization, many innovations brought about by technology could produce chaos and instability. Organizations generally tend to suffer from, and spend too much time reflecting on, their past dilemmas. However, given the recent phenomenon of rapid changes in technology, organizations can no longer afford the luxury of claiming that there is simply too much else to do to be constantly worrying about technology. Indeed, Lounamaa and March (1987) state that organizations can no longer support the claim that too-frequent changes will inhibit learning. The fact is that such changes must be taken as evolutionary, and as a part of the daily challenges facing any organization. Because a learning organization is one that creates structure and strategies, it is positioned to facilitate the learning of all its members, during the ongoing infiltration of technology-driven agents of change. Boland et al. (1994) show that information systems based on multimedia technologies may enhance the appreciation of diverse interpretations within organizations and, as such, support learning organizations. Since learning organizations are deliberately created to facilitate the learning of their members, understanding the urgency of technological changes can provide the stimulus to support planned learning.

Many of the techniques used in the Ravell case study were based on the use of learning organizational techniques, many of which were pioneered by Argyris and Schön (1996). Their work focuses on using "action science" methods to create and maintain learning organizations. A key component of action science is the use of reflective practices—including what is commonly known among researchers and practitioners as reflection in action and reflection on action. *Reflection with action* is the term I use as a rubric for these various methods, involving reflection in relation to activity. Reflection has received a number of definitions, from different sources in the literature. Depending on the emphasis, whether on theory or practice, definitions vary from philosophical articulation (Dewey, 1933; Habermas, 1998), to practice-based formulations, such as Kolb's (1984b) use of reflection in the experiential learning cycle. Specifically, reflection with action carries the resonance of Schön's (1983) twin constructs: reflection on action and reflection in action, which emphasize reflection in retrospect, and reflection to determine which actions to take in the present or immediate future, respectively. Dewey (1933) and Hullfish and Smith (1978) also suggest that the use of reflection supports an implied purpose: individuals reflect for a purpose that leads to the processing of a useful outcome. This formulation suggests the possibility of reflection that is future oriented—what we might call "reflection to action." These are methodological orientations covered by the rubric.

Reflective practices are integral to ROD because so many technology-based projects are event driven and require individuals to reflect before, during, and after actions. Most important to this process is that these reflections are individually driven and that technology projects tend to accelerate the need for rapid decisions. In other words, there are more dynamic decisions to be made in less time. Without operating in the kind of formation that is a learning organization, IT departments cannot maintain the requisite infrastructure to develop products timely on time and support business units—something that clearly is not happening if we look at the existing lateness of IT projects. With respect to the role of reflection in general, the process can be individual or organizational. While groups can reflect, it is in being reflective that individuals bring about "an orientation to their everyday lives," according to Moon (1999). "For others reflection comes about when conditions in the learning environment are appropriate" (p. 186). However, IT departments have long suffered from not having the conditions to support such an individual learning environment. This is why implementing a learning organization is so appealing as a remedy for a chronic problem.

Communities of Practice

Communities of practice are based on the assumption that learning starts with engagement in social practice and that this practice is the fundamental construct by which individuals learn (Wenger, 1998). Thus, communities of practice are formed to get things done by using a shared way of pursuing interest. For individuals, this means that learning is a way of engaging in, and contributing to, the practices of their communities. For specific communities, on the other hand, it means that learning is a way of refining their distinctive practices and ensuring new generations of members. For entire organizations, it means that learning is an issue of sustaining interconnected communities of practice, which define what an organization knows and contributes to the business. The notion of communities of practice supports the idea that learning is an "inevitable part of participating in social life and practice" (Elkjaer, 1999, p. 75). Communities of practice also include assisting members of the community, with the particular focus on improving their skills. This is also known as *situated learning*. Thus, communities of practice are very much a social learning theory, as opposed to one that is based solely on the individual. Communities of practice have been called *learning in working*, in which learning is an inevitable part of working together in a social setting. Much of this concept implies that learning, in some form or other will occur, and that it is accomplished within a framework of social participation, not solely or simply in the individual mind. In a world that is changing significantly due to technological innovations, we should recognize the need for organizations, communities, and individuals to embrace the complexities of being interconnected at an accelerated pace.

There is much that is useful in the theory of communities of practice and that justifies its use in ROD. While so much of learning technology is event driven and individually learned, it would be shortsighted to believe that it is the only way learning can occur in an organization. Furthermore, the enormity and complexity of technology requires a community focus. This would be especially useful within the confines of specific departments that are in need of understanding how to deal with technological dynamism. That is, preparation for using new technologies cannot be accomplished by waiting for an event to occur. Instead, preparation can be accomplished by creating a community that can assess technologies as a part of the normal activities of an organization. Specifically, this means that, through the infrastructure of a community, individuals can determine how they will organize themselves to operate with emerging technologies, what education they will need, and what potential strategic integration they will need to prepare for changes brought on by technology. Action in this context can be viewed as a continuous process, much in the same way that I have presented technology as an ongoing accelerating variable. However, Elkjaer (1999) argues that the continuous process cannot exist without individual interaction. As he states: "Both individual and collective activities are grounded in the past, the present, and the future. Actions and interactions take place between and among group members and should not be viewed merely as the actions and interactions of individuals" (p. 82).

Based on this perspective, technology can be handled by the actions (community) and interactions (individuals) of the organization as shown in Figure 4.2.

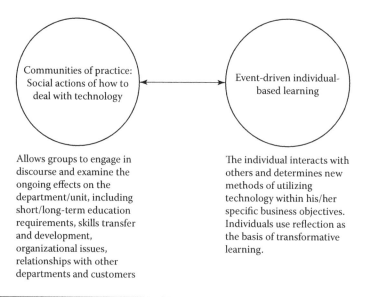

Figure 4.2   Technology relationship between communities and individuals.

It seems logical that communities of practice provide the mechanism to assist, particularly, with the cultural assimilation component of ROD. Indeed, cultural assimilation targets the behavior of the community, and its need to consider what new organizational structures can better support emerging technologies. I have, in many ways, already established and presented the challenge of what should be called the "community of IT practice" and its need to understand how to restructure to meet the needs of the organization. This is the kind of issue that does not lend itself to event-driven, individual learning, but rather to a more community-based process that can deal with the realignment of departmental relationships.

Essentially, communities of IT practice must allow for the continuous evolution of learning based on emergent strategies. Emergent strategies acknowledge unplanned action. Such strategies are defined as patterns that develop in the absence of intentions (Mintzberg & Waters, 1985). Emergent strategies can be used to gather groups that can focus on issues not based on previous plans. These strategies can be thought of as creative approaches to proactive actions. Indeed, a frustrating aspect of technology is its uncertainty. Ideas and concepts borrowed from communities of practice can help departments deal with the evolutionary aspects of technological dynamism.

The relationship, then, between communities of practice and technology is significant. Many of the projects involving IT have been traditionally based on informal processes of learning. While there have been a number of attempts to computerize knowledge using various information databases, they have had mixed results. A "structured" approach to creating knowledge reporting is typically difficult to establish and maintain. Many IT departments have utilized International Organization for Standardization (ISO) 9000 concepts. The ISO is a worldwide organization that defines quality processes through formal structures. It attempts to take knowledge-based information and transfer it into specific and documented steps that can be evaluated as they occur. Unfortunately, the ISO 9000 approach, even if realized, is challenging when such knowledge and procedures are undergoing constant and unpredictable change. Technological dynamism creates too many uncertainties to be handled by the extant discourses on how organizations have dealt with change variables. Communities of practice provide an umbrella of discourses that are necessary to deal with ongoing and unpredictable interactions established by emerging technologies.

Support for this position is found in the fact that technology requires accumulative collective learning that needs to be tied to social practices; this way, project plans can be based on learning as a participatory act. One of the major advantages of communities of practice is that they can integrate key competencies into the very fabric of the organization (Lesser et al., 2000). The typical disadvantage of IT is that its staff needs to serve multiple organizational structures simultaneously. This requires that priorities be set by the organization. Unfortunately, it is difficult, if not impossible, for IT departments to establish such priorities without engaging in concepts of communities of practice that allow for a more integrated process of negotiation and determination. Much of the process of communities of practice would be initiated by strategic integration and result in many cultural assimilation changes; that is, the process of implementing communities of practice will necessitate changes in cultural behavior and organization processes.

As stated, communities-of-practice activities can be initiated via the strategic integration component of ROD. According to Lesser et al. (2000), a knowledge strategy based on communities of practice consists of seven basic steps (Table 4.1).

Lesser and Wenger (2000) suggest that communities of practice are heavily reliant on innovation: "Some strategies rely more on innovation than others for their success. ... Once dependence on innovation needs have been clarified, you can work to create new knowledge where innovation matters" (p. 8). Indeed, electronic communities of practice are different from physical communities. IT provides another dimension to how technology affects organizational learning. It does so by creating new ways in which communities of practice operate. In the complexity of ways that it affects us, technology has a dichotomous relationship with communities of practice. That is, there is a two-sided issue: (1) the need for communities of practice to implement IT projects and integrate them better into learning organizations, and (2) the expansion of electronic communities of practice invoked by technology, which can, in turn, assist in organizational learning, globally and culturally.

The latter issue establishes the fact that a person can now readily be a member of many electronic communities, and in many different

Table 4.1 Extended Seven Steps of Community of Practice Strategy

STEP	COMMUNITIES-OF-PRACTICE STEP	TECHNOLOGY EXTENSION
1	Understanding strategic knowledge needs: What knowledge is critical to success.	Understanding how technology affects strategic knowledge, and what specific technological knowledge is critical to success.
2	Engaging practice domains: People form communities of practice to engage in and identify with.	Technology identifies groups, based on business-related benefits; requires domains to work together toward measurable results.
3	Developing communities: How to help key communities reach their full potential.	Technologies have life cycles that require communities to continue; treats the life cycle as a supporter for attaining maturation and full potential.
4	Working the boundaries: How to link communities to form broader learning systems.	Technology life cycles require new boundaries to be formed. This will link other communities that were previously outside discussions and thus, expand input into technology innovations.
5	Fostering a sense of belonging: How to engage people's identities and sense of belonging.	The process of integrating communities: IT and other organizational units will create new evolving cultures that foster belonging as well as new social identities.
6	Running the business: How to integrate communities of practice into running the business of the organization.	Cultural assimilation provides new organizational structures that are necessary to operate communities of practice and to support new technological innovations.
7	Applying, assessing, reflecting, renewing: How to deploy knowledge strategy through waves of organizational transformation.	The active process of dealing with multiple new technologies that accelerates the deployment of knowledge strategy. Emerging technologies increase the need for organizational transformation.

capacities. Electronic communities are different, in that they can have memberships that are short-lived and transient, forming and re-forming according to interest, particular tasks, or commonality of issue. Communities of practice themselves are utilizing technologies to form multiple and simultaneous relationships. Furthermore, the growth of international communities resulting from ever-expanding global economies has created further complexities and dilemmas.

Thus far, I have presented communities of practice as an infrastructure that can foster the development of organizational learning to support the existence of technological dynamism. Most of what I presented has an impact on the cultural assimilation component of ROD—that is, affecting organizational structure and the way things need to be done. However, technology, particularly the strategic integration component of ROD, fosters a more expanded vision of what can represent a community of practice. What does this mean? Communities of practice, through the advent of strategic integration, have expanded to include electronic communities. While technology can provide organizations with vast electronic libraries that end up as storehouses of information, they are only valuable if they are allowed to be shared within the community. Although IT has led many companies to imagine a new world of leveraged knowledge, communities have discovered that just storing information does not provide for effective and efficient use of knowledge. As a result, many companies have created these "electronic" communities so that knowledge can be leveraged, especially across cultures and geographic boundaries. These electronic communities are predictably more dynamic as a result of what technology provides to them. The following are examples of what these communities provide to organizations:

- Transcending boundaries and exchanging knowledge with internal and external communities. In this circumstance, communities are extending not only across business units, but also into communities among various clients—as we see developing in advanced e-business strategies. Using the Internet and intranets, communities can foster dynamic integration of the client, an important participant in competitive advantage. However, the expansion of an external community, due to emergent electronics, creates yet another need for the implementation of ROD.
- Creating "Internet" or electronic communities as sources of knowledge (Teigland, 2000), particularly for technical-oriented employees. These employees are said to form "communities of techies": technical participants, composed largely of the IT staff, who have accelerated means to come into contact with business-related issues. In the case of Ravell, I created small communities by moving IT staff to allow them to experience the user's need; this move is directly related to the larger, and expanded, ability of using electronic communities of practice.

- Connecting social and workplace communities through sophisticated networks. This issue links well to the entire expansion of issues surrounding organizational learning, in particular, learning organization formation. It enfolds both the process and the social dialectic issues so important to creating well-balanced communities of practice that deal with organizational-level and individual development.

- Integrating teleworkers and non-teleworkers, including the study of gender and cultural differences. The growth of distance workers will most likely increase with the maturation of technological connectivity. Videoconferencing and improved media interaction through expanded broadband will support further developments in virtual workplaces. Gender and culture will continue to become important issues in the expansion of existing models that are currently limited to specific types of workplace issues. Thus, technology allows for the "globalization" of organizational learning needs, especially due to the effects of technological dynamism.

- Assisting in computer-mediated communities. Such mediation allows for the management of interaction among communities, of who mediates their communications criteria, and of who is ultimately responsible for the mediation of issues. Mature communities of practice will pursue self-mediation.

- Creating "flame" communities. A *flame* is defined as a lengthy, often personally insulting, debate in an electronic community that provides both positive and negative consequences. Difference can be linked to strengthening the identification of common values within a community but requires organizational maturation that relies more on computerized communication to improve interpersonal and social factors to avoid miscommunications (Franco et al., 2000).

- Storing collective knowledge in large-scale libraries and databases. As Einstein stated: "Knowledge is experience. Everything else is just information." Repositories of information are not knowledge, and they often inhibit organizations from sharing important knowledge building blocks that affect technical, social, managerial, and personal developments that are critical for learning organizations (McDermott, 2000).

Ultimately, these communities of practice are forming new social networks, which have established the cornerstone of "global connectivity, virtual communities, and computer-supported cooperative work" (Wellman et al., 2000, p. 179). These social networks are creating new cultural assimilation issues, changing the very nature of the way organizations deal with and use technology to change how knowledge develops and is used via communities of practice. It is not, therefore, that communities of practice are new infrastructure or social forces; rather, the difference is in the way they communicate. Strategic integration forces new networks of communication to occur (the IT effect on communities of practice), and the cultural assimilation component requires communities of practice to focus on how emerging technologies are to be adopted and used within the organization.

In sum, what we are finding is that technology creates the need for new organizations that establish communities of practice. New members enter the community and help shape its cognitive schemata. Aldrich (2001) defines *cognitive schemata* as the "structure that represents organized knowledge about persons, roles, and events" (p. 148). This is a significant construct in that it promotes the importance of a balanced evolutionary behavior among these three areas. Rapid learning, or organizational knowledge, brought on by technological innovations can actually lessen progress because it can produce premature closure (March, 1991). Thus, members emerge out of communities of practice that develop around organizational tasks. They are driven by technological innovation and need constructs to avoid premature closure, as well as ongoing evaluation of perceived versus actual realities. As Brown and Duguid (1991, p. 40) state:

> The complex of contradictory forces that put an organization's assumptions and core beliefs in direct conflict with members' working, learning, and innovating arises from a thorough misunderstanding of what working, learning, and innovating are. As a result of such misunderstandings, many modern processes and technologies, particularly those designed to downskill, threaten the robust working, learning, and innovating communities and practice of the workplace.

This perspective can be historically justified. We have seen time and time again how a technology's original intention is not realized yet still productive. For instance, many uses of e-mail by individuals were hard to predict. It may be indeed difficult, if not impossible, to predict the eventual impact of a technology on an organization and provide competitive advantages. However, based on evolutionary theories, it may be beneficial to allow technologies to progress from driver-to-supporter activity. Specifically, this means that communities of practice can provide the infrastructure to support growth from individual-centered learning; that is, to a less event-driven process that can foster systems thinking, especially at the management levels of the organization. As organizations evolve into what Aldrich (2001) call "bounded entities," interaction behind boundaries heightens the salience of cultural difference. Aldrich's analysis of knowledge creation is consistent with what he called an "adaptive organization"—one that is goal oriented and learns from trial and error (individual-based learning)—and a "knowledge development" organization (system-level learning). The latter consists of a set of interdependent members who share patterns of belief. Such an organization uses inferential and vicarious learning and generates new knowledge from both experimentation and creativity. Specifically, learning involves sense making and builds on the knowledge development of its members. This becomes critical to ROD, especially in dealing with change driven by technological innovations. The advantages and challenges of virtual teams and communities of practice are expanded in Chapter 7, in which I integrate the discussion with the complexities of outsourcing teams.

Learning Preferences and Experiential Learning

The previous sections of this chapter focused on organizational learning, particularly two component theories and methods: learning organizations and communities of practice. Within these two methods, I also addressed the approaches to learning; that is, learning that occurs on the individual and the organizational levels. I advocated the position that both system and individual learning need to be part of the equation that allows a firm to attain ROD. Notwithstanding how and when system and individual learning occurs, the investigation of how individuals learn must be a fundamental part of any theory-to-practice effort, such as the present one. Indeed, whether one favors a view of learning as occurring on the organizational or on the individual level (and it occurs on both), we have to recognize that individuals are, ultimately, those who must continue to learn. Dewey (1933) first explored the concepts and values of what he called "experiential learning." This type of learning comes from the experiences that adults have accrued over the course of their individual lives. These experiences provide rich and valuable forms of "literacy," which must be recognized as important components to overall learning development. Kolb (1984a) furthered Dewey's research and developed an instrument that measures individual preferences or styles in which adults learn, and how they respond to day-to-day scenarios and concepts. Kolb's (1999) Learning Style Inventory (LSI) instrument allows adults to better understand how they learn. It helps them understand how to solve problems, work in teams, manage conflicts, make better career choices, and negotiate personal and professional relationships. Kolb's research provided a basis for comprehending the different ways in which adults prefer to learn, and it elaborated the distinct advantages of becoming a balanced learner.

The instrument schematizes learning preferences and styles into four quadrants: *concrete experience*, *reflective observation*, *abstract conceptualization*, and *active experimentation*. Adults who prefer to learn through concrete experience are those who need to learn through actual experience, or compare a situation with reality. In reflective observation, adults prefer to learn by observing others, the world around them, and what they read. These individuals excel in group discussions and can effectively reflect on what they see and read. Abstract conceptualization refers to learning, based on the assimilation of facts and information presented, and read. Those who prefer to learn by active experimentation do so through a process of evaluating consequences; they learn by examining the impact of experimental situations. For any individual, these learning styles often work in combinations. After classifying an individual's responses to questions, Kolb's instrument determines the nature of these combinations. For example, an individual can have a learning style in which he or she prefers to learn from concrete experiences using reflective observation as opposed to actually "doing" the activity. Figure 4.3 shows Kolb's model in the form of a "learning wheel." The wheel graphically shows

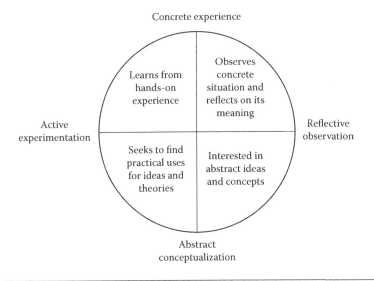

Figure 4.3   Kolb's Learning Style Inventory.

an individual's learning style inventory, reflecting a person's strengths and weaknesses with respect to each learning style.

Kolb's research suggests that learners who are less constrained by learning preferences within a distinct style are more balanced and are better learners because they have available to them more dimensions in which to learn. This is a significant concept; it suggests that adults who have strong preferences may not be able to learn when faced with learning environments that do not fit their specific preference. For example, an adult who prefers group discussion and enjoys reflective conversation with others may feel uncomfortable in a less interpersonal, traditional teaching environment. The importance of Kolb's LSI is that it helps adults become aware that such preferences exist.

McCarthy's (1999) research furthers Kolb's work by investigating the relationship between learning preferences and curriculum development. Her Learning Type Measure (4Mat) instrument mirrors and extends the Kolb style quadrants by expressing preferences from an individual's perspective on how to best achieve learning. Another important contribution in McCarthy's extension of Kolb's work is the inclusion of brain function considerations, particularly in terms of hemisphericity. McCarthy focuses on the cognitive functions associated with the right hemisphere (perception) and left hemisphere (process) of the brain. Her 4Mat system shows how adults, in each style quadrant, perceive learning with the left hemisphere of the brain and how it is related to processing in the right hemisphere. For example, for Type 1 learners (concrete experience and reflective observation), adults perceive in a concrete way and process in a reflective way. In other words, these adults prefer to learn by actually doing a task and then processing the experience by reflecting on what they experienced during the task. Type 2 learners (reflective observation and abstract conceptualization), however, perceive a task by abstract thinking and process it by developing concepts and theories from their initial ideas. Figure 4.4 shows McCarthy's rendition of the Kolb learning wheel.

The practical claim to make here is that practitioners who acquire an understanding of the concepts of the experiential learning models will be better able to assist individuals in understanding how they learn, how to use their learning preferences during times of

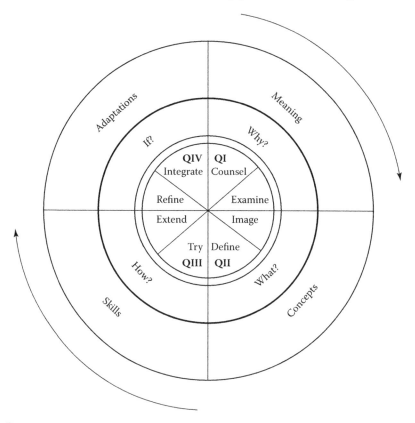

Figure 4.4   McCarthy rendition of the Kolb Learning Wheel.

transition, and the importance of developing other dimensions of learning. The last is particularly useful in developing expertise in learning from individual reflective practices, learning as a group in communities of practice, and participating in both individual transformative learning, and organizational transformations. How, then, does experiential learning operate within the framework of organizational learning and technology? This is shown Figure 4.5 in a combined wheel, called the *applied individual learning for technology model,* which creates a conceptual framework for linking the technology life cycle with organizational learning and experiential learning constructs.

Figure 4.5 expands the wheel into two other dimensions. The first quadrant (QI) represents the feasibility stage of technology. It requires communities to work together, to ascertain why a particular technology might be attractive to the organization. This quadrant is

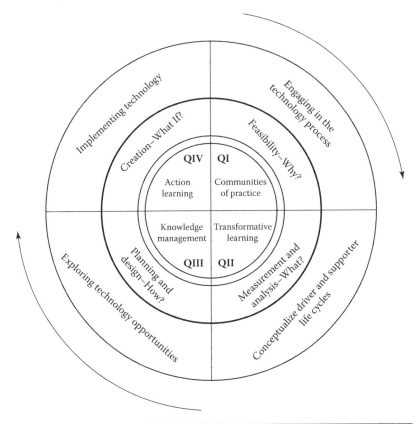

Figure 4.5 Combined applied learning wheel.

best represented by individuals who engage in group discussions to make better connections from their own experiences. The process of determining whether a technology is feasible requires integrated discourse among affected communities, who then can make better decisions, as opposed to centralized or individual and predetermined decisions on whether to use a specific technology. During this phase, individuals need to operate in communities of practice, as the infrastructure with which to support a democratic process of consensus building.

The second quadrant (QII) corresponds to measurement and analysis. This operation requires individuals to engage in specific details to determine and conceptualize driver and supporter life cycles analytically. Individuals need to examine the specific details to understand "what" the technology can do, and to reflect on what it means to them, and their business unit. This analysis is measured with respect to what the ROI will be, and which driver and supporter functions will be used. This process requires transformation theory that allows individuals to perceive and conceptualize which components of the technology can transform the organization.

Quadrant 3 (QIII), design and planning, defines the "how" component of the technology life cycle. This process involves exploring technology opportunities after measurement and analysis have been completed. The process of determining potential uses for technology requires knowledge of the organization. Specifically, it needs the abstract concepts developed in QII to be integrated with tacit knowledge, to then determine possible applications where the technology can succeed. Thus, knowledge management becomes the predominant mechanism for translating what has been conceptualized into something explicit (discussed further in Chapter 5).

Quadrant 4 (QIV) represents the implementation-and-creation step in the technology life cycle. It addresses the hypothetical question of "What if?" This process represents the actual implementation of the technology. Individuals need to engage in action learning techniques, particularly those of reflective practices. The implementation step in the technology life cycle is heavily dependent on the individual. Although there are levels of project management, the essential aspects of what goes on inside the project very much relies on the individual performances of the workers.

Social Discourse and the Use of Language

The successful implementation of communities of practice fosters heavy dependence on social structures. Indeed, without understanding how social discourse and language behave, creating and sustaining the internal interactions within and among communities of practice are not possible. In taking individuals as the central component for continued learning and change in organizations, it becomes important to work with development theories that can measure and support individual growth and can promote maturation with the promotion of organizational/system thinking (Watkins & Marsick, 1993). Thus, the basis for establishing a technology-driven world requires the inclusion of linear and circular ways of promoting learning. While there is much that we will use from reflective action concepts designed by Argyris and Schön (1996), it is also crucial to incorporate other theories, such as marginality, transitions, and individual development.

Senge (1990) also compares learning organizations with engineering innovation; he calls these engineering innovations "technologies." However, he also relates innovation to human behavior and distinguishes it as a "discipline." He defines *discipline* as "a body of theory and technique that must be studied and mastered to be put into practice, as opposed to an enforced order or means of punishment" (p. 10). A discipline, according to Senge, is a developmental path for acquiring certain skills or competencies. He maintains the concept that certain individuals have an innate "gift"; however, anyone can develop proficiency through practice. To practice a discipline is a lifelong learning process—in contrast to the work of a learning organization. Practicing a discipline is different from emulating a model. This book attempts to bring the arenas of discipline and technology into some form of harmony. What technology offers is a way of addressing the differences that Senge proclaims in his work. Perhaps this is what is so interesting and challenging about attempting to apply and understand the complexities of how technology, as an engineering innovation, affects the learning organization discipline—and thereby creates a new genre of practices. After all, I am not sure that one can master technology as either an engineering component, or a discipline.

Technology dynamism and ROD expand the context of the globalizing forces that have added to the complexity of analyzing "the language and symbolic media we employ to describe, represent, interpret, and theorize what we take to be the facticity of organizational life" (Grant et al., 1998, p. 1). ROD needs to create what I call the "language of technology." How do we then incorporate technology in the process of organizing discourse, or how has technology affected that process? We know that the concept of discourse includes language, talk, stories, and conversations, as well as the very heart of social life, in general. Organizational discourse goes beyond what is just spoken; it includes written text and other informal ways of communication. Unfortunately, the study of discourse is seen as being less valuable than action. Indeed, discourse is seen as a passive activity, while "doing" is seen as supporting more tangible outcomes. However, technology has increased the importance of sensemaking media as a means of constructing and understanding organizational identities. In particular, technology, specifically the use of e-mail, has added to the instability of language, and the ambiguities associated with metaphorical analysis—that is, meaning making from language as it affects organizational behavior. Another way of looking at this issue is to study the metaphor, as well as the discourse, of technology. Technology is actually less understood today, a situation that creates even greater reason than before for understanding its metaphorical status in organizational discourse—particularly with respect to how technology uses are interpreted by communities of practice. This is best shown using the schema of Grant et al. of the relationship between content and activity and how, through identity, skills, and emotion, it leads to action (Figure 4.6).

To best understand Figure 4.4 and its application to technology, it is necessary to understand the links between talk and action. It is the activity and content of conversations that discursively produce identities, skills, and emotions, which in turn lead to action. Talk, in respect to conversation and content, implies both oral and written forms of communications, discourse, and language. The written aspect can obviously include technologically fostered communications over the Internet. It is then important to examine the unique conditions that technology brings to talk and its corresponding actions.

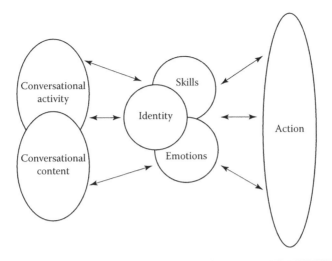

Figure 4.6   Grant's schema—relationship between content and activity.

Identity

Individual identities are established in collaborations on a team, or in being a member of some business committee. Much of the theory of identity development is related to how individuals see themselves, particularly within the community in which they operate. Thus, how active or inactive we are within our communities, shapes how we see ourselves and how we deal with conversational activity and content. Empowerment is also an important part of identity. Indeed, being excluded or unsupported within a community establishes a different identity from other members of the group and often leads to marginality (Schlossberg, 1989).

Identities are not only individual but also collective, which to a large extent contributes to cultures of practice within organizational factions. It is through common membership that a collective identity can emerge. Identity with the group is critical during discussions regarding emerging technologies and determining how they affect the organization. The empowerment of individuals, and the creation of a collective identity, are therefore important in fostering timely actions that have a consensus among the involved community.

Skills

According to Hardy et al. (1998, p. 71), conversations are "arenas in which particular skills are invested with meaning." Watson (1995) suggests that conversations not only help individuals acquire "technical skills" but also help develop other skills, such as being persuasive. Conversations that are about technology can often be skewed toward the recognition of those individuals who are most "technologically talented." This can be a problem when discourse is limited to who has the best "credentials" and can often lead to the undervaluing of social production of valued skills, which can affect decisions that lead to actions.

Emotion

Given that technology is viewed as a logical and rational field, the application of emotion is not often considered a factor of action. Fineman (1996) defines *emotion* as "personal displays of affected, or 'moved' and 'agitated' states—such as joy, love, fear, anger, sadness, shame, embarrassment,"—and points out that these states are socially constructed phenomena. There is a positive contribution from emotional energy as well as a negative one. The consideration of positive emotion in the organizational context is important because it drives action (Hardy et al., 1998). Indeed, action is more emotion than rational calculation. Unfortunately, the study of emotions often focuses on its negative aspects. Emotion, however, is an important part of how action is established and carried out, and therefore warrants attention in ROD.

Identity, skills, and emotion are important factors in how talk actually leads to action. Theories that foster discourse, and its use in organizations, on the other hand, are built on linear paths of talk and action. That is, talk can lead to action in a number of predefined paths. Indeed, talk is typically viewed as "cheap" without action or, as is often said, "action is valued," or "action speaks louder than words." Talk, from this perspective, constitutes the dynamism of what must occur with action science, communities of practice, transformative learning, and, eventually, knowledge creation and management. Action, by contrast, can be viewed as the measurable outcomes that have been eluding organizational learning scholars. However, not all actions lead to measurable outcomes. Marshak (1998) established three types of talk that lead to action: *tool-talk*, *frame-talk*, and *mythopoetic-talk*:

1. *Tool-talk* includes "instrumental communities required to: discuss, conclude, act, and evaluate outcomes" (p. 82). What is most important in its application is that tool-talk be used to deal with specific issues for an identified purpose.
2. *Frame-talk* focuses on interpretation to evaluate the meanings of talk. Using frame-talk results in enabling implicit and explicit assessments, which include symbolic, conscious, preconscious, and contextually subjective dimensions.
3. *Mythopoetic-talk* communicates ideogenic ideas and images (i.e., myths and cosmologies) that can be used to communicate the nature of how to apply tool-talk and frame-talk within the particular culture or society. This type of talk allows for concepts of intuition and ideas for concrete application.

Furthermore, it has been shown that organizational members experience a difficult and ambiguous relationship, between discourse that makes sense, and non-sense—what is also known as "the struggle with sense" (Grant et al., 1998). There are two parts that comprise non-sense: The first is in the difficulties that individuals experience in understanding why things occur in organizations, particularly when their actions "make no sense." Much of this difficulty can be correlated with political issues that create "nonlearning" organizations. However, the second condition of non-sense is more applicable, and more important, to the study of ROD than the first—that is, non-sense associated with acceleration in the organizational change process. This area comes from the taken-for-granted assumptions about the realities of how the organization operates, as opposed to how it can operate. Studies performed by Wallemacq and Sims (1998) provide examples of how organizational interventions can decompose stories about non-sense and replace them with new stories that better address a new situation and can make sense of why change is needed. This phenomenon is critical to changes established, or responded to, by the advent of new technologies. Indeed, technology has many nonsensical or false generalizations regarding how long it takes to implement a product, what might be the expected outcomes, and so on. Given

the need for ROD—due to the advent of technology—there is a concomitant need to reexamine "old stories" so that the necessary change
agents can be assessed and put into practice. Ultimately, the challenge
set forth by Wallemacq and Sims is especially relevant, and critical,
since the very definition of ROD suggests that communities need
to accelerate the creation of new stories—stories that will occur at
unpredictable intervals. Thus, the link between discourse, organizational learning, and technology is critical to providing ways in which
to deal with individuals and organizations facing the challenge of
changing and evolving.

Grant's (1996) research shows that sense making using media and
stories provided effective ways of constructing and understanding
organizational identities. Technology affects discourse in a similar
way that it affects communities of practice; that is, it is a variable that
affects the way discourse is used for organizational evolution. It also
provides new vehicles on how such discourse can occur. However, it is
important not to limit discourse analysis to merely being about "texts,"
emotion, stories, or conversations in organizations. Discourse analysis
examines "the constructing, situating, facilitating, and communicating of diverse cultural, instrumental, political, and socio-economic
parameters of 'organizational being'" (Grant, 1996, p. 12). Hence,
discourse is the essential component of every organizational learning effort. Technology accelerates the need for such discourse, and
language, in becoming a more important part of the learning maturation process, especially in relation to "system" thinking and learning.
I propose then, as part of a move toward ROD, that discourse theories
must be integrated with technological innovation and be part of the
maturation in technology and in organizational learning.

The overarching question is how to apply these theories of discourse and language to learning within the ROD framework and paradigm. First, let us consider the containers of types of talk discussed
by Marshak (1998) as shown in Figure 4.7.

These types of talk can be mapped onto the technology wheel, so that
the most appropriate oral and written behaviors can be set forth within
each quadrant, and development life cycle, as shown in Figure 4.8.

Mythopoetic-talk is most appropriate in Quadrant 1 (QI), where
the fundamental ideas and issues can be discussed in communities of
practice. These technological ideas and concepts, deemed feasible, are

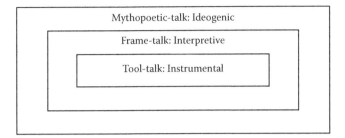

Figure 4.7 Marshak's type of talk containers.

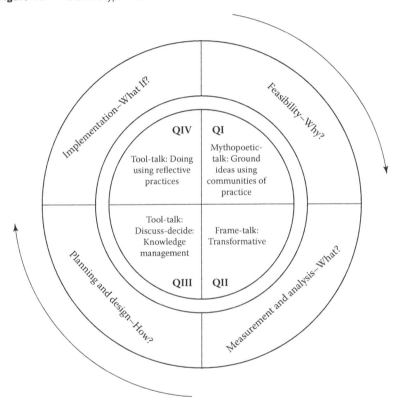

Figure 4.8 Marshak's model mapped to the technology learning wheel.

then analyzed through frame-talk, by which the technology can be evaluated in terms of how it meets the fundamental premises established in QI. Frame-talk also reinforces the conceptual legitimacy of how technology will transform the organization while providing appropriate ROI. Tool-talk represents the process of identifying applications and actually implementing them. For this reason, tool-talk exists in both QIII and QIV. The former quadrant represents the discussion-to-decision portion, and the latter represents the actual doing and completion of the project itself. In QIII, table-talk requires knowledge management to transition technology concepts into real options. QIV transforms these real options into actual projects, in which, reflecting on actual practices during implementation, provides an opportunity for individual- and organizational-level learning.

Marshak's (1998) concept of containers and cycles of talk and action are adapted and integrated with cyclical and linear maturity models of learning. However, discourse and language must be linked to performance, which is why it needs to be part of the discourse and language-learning wheel. By integrating discourse and language into the wheel, individual and group activities can use discourse and language as part of reflective practices to create an environment that can foster action that leads to measurable outcomes. This process, as explained throughout this book, is of paramount importance in understanding how discourse operates with ROD in the information age.

Linear Development in Learning Approaches

Focusing only on the role of the individual in the company is an incomplete approach to formulating an effective learning program. There is another dimension to consider that is based on learning maturation. That is, where in the life cycle of learning are the individuals and the organization? The best explanation of this concept is the learning maturation experience at Ravell. During my initial consultation at Ravell, the organization was at a very early stage of organizational learning. This was evidenced by the dependence of the organization on event-driven and individual reflective practice learning. Technology acted as an accelerator of learning—it required IT to design a new network during the relocation of the company. Specifically, the acceleration, operationalized by a physical move, required IT to establish new relationships with line management. The initial case study concluded that there was a cultural change as a result of these new relationships—cultural assimilation started to occur using organizational learning techniques, specifically reflective practices.

After I left Ravell, another phase in the evolution of the company took place. A new IT director was hired in my stead, who attempted to reinstate the old culture: centralized infrastructure, stated operational boundaries, and separations that mandated anti-learning organizational behaviors. After six months, the line managers, faced with having to revert back to a former operating culture, revolted and demanded the removal of the IT director. This outcome, regrettable as it may be, is critical in proving the conclusion of the original study that the culture at Ravell had indeed evolved from its state, at the time of my arrival. The following are two concrete examples that support this notion:

1. The attempt of the new IT director to "roll back" the process to a former cultural state was unsuccessful, showing that a new evolving culture had indeed occurred.
2. Line managers came together from the established learning organization to deliver a concerted message to the executive team. Much of their learning had now shifted to a social organization level that was based less on events and was more holistic with respect to the goals and objectives of the organization.

Thus, we see a shift from an individual-based learning process to one that is based more on the social and organizational issues to stimulate transformation. This transformation in learning method occurred within the same management team, suggesting that changes in learning do occur over time and from experience. Another way of viewing the phenomenon is to see Ravell as reaching the next level of organizational learning or maturation with learning. Consistent with the conclusion of the original study, technology served to accelerate the process of change or accelerate the maturation process of organizational learning.

Another phase (Phase II) of Ravell transpired after I returned to the company. I determined at that time that the IT department needed to be integrated with another technology-based part of the business—the unit responsible for media and engineering services (as opposed to IT). While I had suggested this combination eight months earlier, the organization had not reached the learning maturation to understand why such a combination was beneficial. Much of the reason it did not occur earlier, can also be attributed to the organization's inability to manage ROD, which, if implemented, would have made the integration more obvious. The initial Ravell study served to bring forth the challenges of cultural assimilation, to the extent that the organization needed to reorganize itself and change its behavior. In phase II, the learning process matured by accelerating the need for structural change in the actual reporting processes of IT.

A year later, yet another learning maturation phase (phase III) occurred. In Ravell, Phase III, the next stage of learning maturation, allowed the firm to better manage ROD. After completing the merger of the two technically related business units discussed (phase II), it became necessary to move a core database department completely out of the combined technology department, and to integrate it with a business unit. The reason for this change was compelling and brought to light a shortfall in my conclusions from the initial study. It appears that as organizational learning matures within ROD, there is an increasing need to educate the executive management team of the organization. This was not the case during the early stages of the case study. The limitation of my work, then, was that I predominantly interfaced with line management and neglected to include executives in the learning. During that time, results were encouraging, so there was little reason for me to include executives in event-driven issues, as discussed. Unfortunately, lacking their participation fostered a disconnection with the strategic integration component of ROD. Not participating in ROD created executive ignorance of the importance that IT had on the strategy of the business. Their lack of knowledge resulted in chronic problems with understanding the relationship and value of IT on the business units of the organization. This shortcoming resulted in continued conflicts over investments in the IT organization. It ultimately left IT with the inability to defend many of its cost requirements. As stated, during times of economic downturns, firms tend to reduce support organizations. In other words, executive management did not understand the driver component of IT.

After the move of the cohort of database developers to a formal business line unit, the driver components of the group provided the dialogue and support necessary to educate executives. However, this education did not occur based on events, but rather, on using the social and group dynamics of organizational learning. We see here another aspect of how organizational and individual learning methods work together, but evolve in a specific way, as summarized in Table 4.2.

Another way of representing the relationship between individual and organizational learning over time is to chart a "maturity" arc to illustrate the evolutionary life cycle of technology and organizational learning. I call this arc the ROD arc. The arc is designed to assess individual development in four distinct sectors of ROD, each in relation to five developmental stages of organizational learning. Thus, each sector of ROD can be measured in a linear and integrated way. Each stage in the course of the learning development

Table 4.2   Analysis of Ravell's Maturation with Technology

LEARNING	PHASE I	PHASE II	PHASE III
Type of learning	Individual reflective practices used to establish operations and line management.	Line managers defend new culture and participate in less event-driven learning.	Movement away from holistic formation of IT, into separate driver and supporter attributes. Learning approaches are integrated using both individual and organizational methods, and are based on functionality as opposed to being organizationally specific.
Learning outcomes	Early stage of learning organization development.	Combination of event-driven and early-stage social organizational learning formation.	Movement toward social-based organizational decision making, relative to the different uses of technology.
Responsive organizational dynamism: cultural assimilation.	Established new culture; no change in organizational structure.	Cultural assimilation stability with existing structures; early phase of IT organizational integration with similar groups.	Mature use of cultural assimilation, based on IT behaviors (drivers and supporters).
Responsive organizational dynamism: Strategic integration.	Limited integration due to lack of executive involvement.	Early stages of value/needs based on similar strategic alignment.	Social structures emphasize strategic integration based on business needs.

of an organization reflects an underlying principle that guides the process of ROD norms and behaviors; specifically, it guides organizations in how they view and use the ROD components available to them.

The arc is a classificatory scheme that identifies progressive stages in the assimilated uses of ROD. It reflects the perspective—paralleling Knefelkamp's (1999) research—that individuals in an organization are able to move through complex levels of thinking, and to develop independence of thought and judgment, as their careers progress within the management structures available to them. Indeed, assimilation to learning at specific levels of operations and management are not necessarily an achievable end but one that fits into the psychological perspective of what productive employees can be taught about ROD adaptability. Figure 4.9 illustrates the two axes of the arc.

The profile of an individual who assimilates the norms of ROD can be characterized in five developmental stages (vertical axis) along four sectors of literacy (horizontal axis). The arc characterizes an individual at a specific level in the organization. At each level, the arc identifies individual maturity with ROD, specifically strategic integration, cultural assimilation, and the type of learning process (i.e., individual vs. organizational). The arc shows how each tier integrates with another, what types of organizational learning theory best apply, and who needs to be the primary driver within the organization. Thus, the arc provides an organizational schema for how each conceptual component of organizational learning applies to each sector of ROD. It also identifies and constructs a path for those individuals who want to advance in organizational rank; that is, it can be used to ascertain an individual's ability to cope with ROD requirements as a precursor for advancement in management. Each position within a sector, or cell, represents a specific stage of development within ROD. Each cell contains specific definitions that can be used to identify developmental stages of ROD and organizational learning maturation. Figure 4.10 represents the ROD arc with its cell definitions. The five stages of the arc are outlined as follows:

Sectors of responsive organizational dynamism	Operational knowledge	Department/unit view as other	Integrated disposition	Stable operations	Organizational leadership
Strategic integration					
Cultural assimilation					
Organizational learning constructs					
Management level					

Figure 4.9 Reflective organizational dynamism arc model.

Sector variable	Operational knowledge	Department/unit view as other	Integrated disposition	Stable operations	Organizational leadership
Strategic integration	Operations personnel understand that technology has an impact on strategic development, particularly on existing processes	Individual beliefs of strategic impact are incomplete; individual needs to incorporate other views within the department or business unit	Recognition that individual and department views must be integrated to be complete and strategically productive for the department/unit	Changes made to processes at the department/unit level formally incorporate emerging technologies	Departmental strategies are propagated and integrated at organization level
Cultural assimilation	View that technology can and will affect the way the organization operates and that it can affect roles and responsibilities	Changes brought forth by technology need to be assimilated into departments and are dependent on how others participate	Understands need for organizational changes; different cultural behavior new structures are seen as viable solutions	Organizational changes are completed and in operation; existence of new or modified employee positions	Department-level organizational changes and cultural evolution are integrated with organization-wide functions and cultures
Organizational learning constructs	Individual-based reflective practice	Small group-based reflective practices	Interactive with both individual and middle management using communities of practice	Interactive between middle management and executives using social discourse methods to promote transformation	Organizational learning at executive level using knowledge management
Management level	Operations	Operation and middle management	Middle management	Middle management and executive	Executive

Figure 4.10 Responsive organizational dynamism arc.

1. *Operational knowledge*: Represents the capacity to learn, conceptualize, and articulate key issues relating to how technology can have an impact on existing processes and organizational structure. Organizational learning is accomplished through individual learning actions, particularly reflective practices. This stage typically is the focus for operations personnel, who are usually focused on their personal perspectives of how technology affects their daily activities.

2. *Department/unit view as other*: Indicates the ability to integrate points of view about using technology from diverse individuals within the department or business unit. Using these new perspectives, the individual is in position to augment his or her understanding of technology and relate it to others within the unit. Operations personnel participate in small-group learning activities, using reflective practices. Lower levels of middle managers participate in organizational learning that is in transition, from purely individual to group-level thinking.

3. *Integrated disposition*: Recognizes that individual and departmental views on using technology need to be integrated to form effective business unit objectives. Understanding that organizational and cultural shifts need to include all member perspectives, before formulating departmental decisions, organizational learning is integrated with middle managers, using communities of practice at the department level.

4. *Stable operations*: Develops in relation to competence in sectors of ROD appropriate for performing job duties for emerging technologies, not merely adequately, but competitively, with peers and higher-ranking employees in the organization. Organizational learning occurs at the organizational level and uses forms of social discourse to support organizational transformation.

5. *Organizational leadership*: Ability to apply sectors of ROD to multiple aspects of the organization. Department concepts can be propagated to organizational levels, including strategic and cultural shifts, relating to technology opportunities. Organizational learning occurs using methods of knowledge management with executive support. Individuals use their

technology knowledge for creative purposes. They are willing to take risks using critical discernment and what Heath
(1968) calls "freed" decision making.

The ROD arc addresses both individual and organizational
learning. There are aspects of Senge's (1990) "organizational"
approach that are important and applicable to this model. I
have mentioned its appropriateness in regard to the level of the
manager—suggesting that the more senior manager is better positioned to deal with nonevent learning practices. However, there is
yet another dimension within each stage of matured learning. This
dimension pertains to timing. The timing dimension focuses on
a multiple-phase approach to maturing individual and organizational learning approaches. The multiple phasing of this approach
suggests a maturing or evolutionary learning cycle that occurs
over time, in which individual learning fosters the need and the
acceptance of organizational learning methods. This process can
be applied within multiple tiers of management and across different business units.

The ROD arc can also be integrated with the applied individual
learning wheel. The combined models show the individual's cycle of
learning along a path of maturation. This can be graphically shown
to reflect how the wheel turns and moves along the continuum of the
arc (Figure 4.11).

Figure 4.11 shows that an experienced technology learner can
maximize learning by utilizing all four quadrants in each of the
maturity stages. It should be clear that certain quadrants of individual learning are more important to specific stages on the arc.
However, movement through the arc is usually not symmetrical;
that is, individuals do not move equally from stage to stage, within
the dimensions of learning (Langer, 2003). This integrated and
multiphase method uses the applied individual learning wheel
with the arc. At each stage of the arc, an individual will need
to draw on the different types of learning that are available in
the learning wheel. Figure 4.12 provides an example of this concept, which Knefelkamp calls "multiple and simultaneous" (1999),
meaning that learning can take on multiple meanings across different sectors simultaneously.

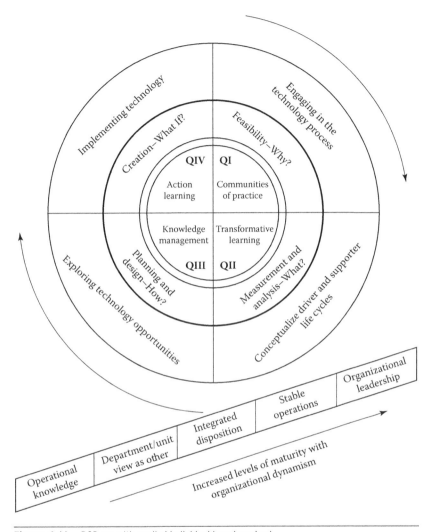

Figure 4.11   ROD arc with applied individual learning wheel.

Figure 4.12 shows that the dimension variables are not necessarily parallel in their linear maturation. This phenomenon is not unusual with linear models, and in fact, is quite normal. However, it also reflects the complexity of how variables mature, and the importance of having the capability and infrastructure to determine how to measure such levels of maturation within dimensions. There are both qualitative and quantitative approaches to this analysis. Qualitative approaches typically include interviewing, ethnographic-type experiences over

Dimension variable	Operational knowledge	Department/unit view as other	Integrated disposition	Stable operations	Organizational leadership
Strategic integration	▓				
Cultural assimilation		▓	▓		
Organizational learning constructs	▓	▓			
Management level		▓			

Figure 4.12 Sample ROD arc.

some predetermined time period, individual journals or diaries, group meetings, and focus groups. Quantitative measures involve the creation of survey-type measures; they are based on statistical results from answering questions that identify the level of maturation of the individual.

The learning models that I elaborate in this chapter are suggestive of the rich complexities surrounding the learning process for individuals, groups, and entire organizations. This chapter establishes a procedure for applying these learning models to technology-specific situations. It demonstrates how to use different phases of the learning process to further mature the ability of an organization to integrate technology strategically and culturally.

5

MANAGING ORGANIZATIONAL LEARNING AND TECHNOLOGY

The Role of Line Management

In Chapter 1, the results of the Ravell case study demonstrated the importance of the role that line managers have, for the success of implementing organizational learning, particularly in the objective of integrating the information technology (IT) department. There has been much debate related to the use of event-driven learning. In particular, there is Senge's (1990) work from his book, *The Fifth Discipline*. While overall, I agree with his theories, I believe that there is a need to critique some of his core concepts and beliefs. That is, Senge tends to make broad generalizations about the limits of event-driven education and learning in organizations. He believes that there is a limitation of learning from experience because it can create limitations to learning based on actions—as he asks: "What happens when we can no longer observe the consequences of our actions?" (Senge, 1990, p. 23).

My research has found that event-driven learning is essential to most workers who have yet to learn through other means. I agree with Senge that not all learning can be obtained through event-oriented thinking, but I feel that much of what occurs at this horizon pertains more to the senior levels than to what many line managers have to deal with as part of their functions in business. Senge's concern with learning methods that focus too much on the individual, perhaps, is more powerful, if we see the learning organization as starting at the top and then working its way down. The position, however, particularly with respect to the integration of technology, is that too much dependence on executive-driven programs to establish and sustain organizational learning, is dangerous. Rather, the line management—or middle managers who fundamentally run the business—is best positioned to make the difference. My hypothesis here is that both top-down and bottom-up approaches to organizational learning are riddled with problems, especially in their ability to sustain outcomes. We cannot be naïve—even our senior executives must drive results to maintain their positions. As such, middle managers, as the key business drivers, must operate in an event- and results-driven world—let us not underestimate the value of producing measurable outcomes, as part of the ongoing growth of the organizational learning practicum.

To explore the role of middle managers further, I draw on the interesting research done by Nonaka and Takeuchi (1995). These researchers examined how Japanese companies manage knowledge creation, by using an approach that they call "middle-up-down." Nonaka and Takeuchi found that middle managers "best communicate the continuous iterative process by which knowledge is created" (p. 127). These middle managers are often seen as leaders of a team, or task, in which a "spiral conversion process" operates and that requires both executive and operations management personnel. Peters and Waterman (1982), among others, often have attacked middle managers as representing a layer of management that creates communication problems and inefficiencies in business processes that resulted in leaving U.S. workers trailing behind their international competitors during the automobile crisis in the 1970s. They advocate a "flattening" of the never-ending levels of bureaucracy responsible for inefficient operations. However, executives often are not aware of details within their operating departments and may not have the ability or time to acquire those details. Operating personnel, on the other hand, do not possess the vision and business aptitudes necessary to establish the kind of knowledge creation that fosters strategic learning.

Middle managers, or what I prefer to identify as line managers (Langer, 2001b), possess an effective combination of skills that can provide positive strategic learning infrastructures. Line managers understand the core issues of productivity in relation to competitive operations and return on investment, and they are much closer to the day-to-day activities that bring forth the realities of how, and when, new strategic processes can be effectively implemented. While many researchers, such as Peters and Waterman, find them to be synonymous with backwardness, stagnation, and resistance to change, middle managers are the core group that can provide the basis for continuous innovation through strategic learning. It is my perspective that the difference of opinion regarding the positive or negative significance middle managers have in relation to organizational learning has to do with the wide-ranging variety of employees who fall into the category of "middle." It strikes me that Peters and Waterman were somewhat on target with respect to a certain population of middle managers, although I would not characterize them as line managers. To justify this position, it is important to clearly establish the differences. Line managers should be defined as pre-executive employees who have reached a position of managing a business unit that contains some degree of return on investment for the business. In effect, I am suggesting that focusing on "middle" managers, as an identifiable group, is too broad. Thus, there is a need to further delineate the different levels of what comprises middle managers, and their roles in the organization.

Line Managers

These individuals usually manage an entire business unit and have "return-on-investment" responsibilities. Line managers should be categorized as those who have middle managers reporting to them; they are, in effect, managers of managers, or, as in some organizations, they serve a "directorial" function. Such individuals are, in many ways, considered future executives and perform many low-end executive tasks. They are, if you will, executives in training. What is significant about this managerial level is the knowledge it carries about operations. However, line managers are still involved in daily operations and maintain their own technical capabilities.

First-Line Managers

First-line individuals manage nonmanagers but can have supervisory employees who report to them. They do not carry the responsibility for a budget line unit but for a department within the unit. These managers have specific goals that can be tied to their performance and to the department's productivity.

Supervisor

A supervisor is the lowest-level middle manager. These individuals manage operational personnel within the department. Their management activities are typically seen as "functions," as opposed to managing an entire operation. These middle managers do not have other supervisors or management-level personnel reporting to them.

We should remember that definitions typically used to characterize the middle sectors of management, as described by researchers like Peters, Nonaka, and others, do not come from exact science. The point must be made that middle managers cannot be categorized by a single definition. The category requires distinctive definitions within each level of stratification presented. Therefore, being more specific about the level of the middle manager can help us determine the manager's role in the strategic learning process. Given that Nonaka and Takeuchi (1995) provide the concept of middle-up-down as it related to knowledge management, I wish to broaden it into a larger subject of strategic learning, as a method of evolving changes in culture and organizational thinking. Furthermore, responsive organizational dynamism (ROD), unlike other organizational studies, represents both situational learning and ongoing evolutionary learning requirements. Evolutionary learning provides a difficult challenge to organizational learning concepts. Evolutionary learning requires significant contribution from middle managers. To understand the complexity of the middle manager, all levels of the organization must be taken into consideration. I call this process *management vectors*.

Management Vectors

Senge's (1990) work addresses some aspects of how technology might affect organizational behavior: "The central message of the Fifth Discipline is more radical than 'radical organization redesign'—namely that our organizations work the way they work, ultimately because of how we think and how we interact" (p. xiv). Technology aspires to be a new variable or catalyst that can change everyday approaches to things—to be the radical change element that forces us to reexamine norms no longer applicable to business operations. On the other hand, technology can be dangerous if perceived unrealistically as a power that possesses new answers to organizational performance and efficiency. In the late 1990s, we experienced the "bust" of the dot-com explosion, an explosion that challenged conventional norms of how businesses operate. Dot-coms sold the concepts that brick-and-mortar operations could no longer compete with new technology-driven businesses and that "older" workers could not be transformed in time to make dot-com organizations competitive. Dot-coms allowed us to depart from our commitment to knowledge workers and learning organizations, which is still true today.

For example, in 2003, IBM at its corporate office in Armonk, New York, laid off 1,000 workers who possessed skills that were no longer perceived as needed or competitive. Rather than retrain workers, IBM determined that hiring new employees to replace them was simply more economically feasible and easier in terms of transforming their organization behaviors. However, in my interview with Stephen McDermott, chief executive officer (CEO) of ICAP Electronic Trading Community (ETC), it became apparent that many of the mystiques of managing technology were incorrect. As he stated, "Managing a technology company is no different from managing other types of businesses." While the technical skills of the IBM workers may no longer be necessary, why did the organization not provide enough opportunities to migrate important knowledge workers to another paradigm of technical and business needs? Widespread worker replacements tell us that few organizational learning infrastructures actually exist. The question is whether technology can provide the stimulus to prompt more organizations to commit to creating infrastructures that support growth and sustained operation. Most important is the question of how we establish infrastructures that can provide the impetus for initial and ongoing learning organizations. This question suggests that the road to working successfully with technology will require the kind of organizational learning that is driven by both individual and organization-wide initiatives. This approach can be best explained by referring to the concept of driver and supporter functions and life cycles of technology presented in Chapter 3. Figure 5.1 graphically shows the relationship between organizational structure and organizational learning needs. We also see that this relationship maps onto driver and supporter functionality.

Figure 5.1 provides an operational overview of the relations between the three general tiers of management in most organizations. These levels or tiers are mapped onto organizational learning approaches; that is, organizational/system or individual. This mapping follows a general view based on what individuals at each of these tiers view or

Management/ operational layers	Driver/support life cycle involvement	Learning approach	Organizational learning method
Executive tier	Driver	Organization system	Knowledge management

⇑

| Middle management tiers | Driver/support life cycle | Organization/ system on driver individual on support | Communities of practice (driver) reflective practices (supporter) |

⇓

| Operations tier | Support | Event-driven individual | Reflective practices |

Figure 5.1   Three-tier organizational structure.

seek as their job responsibilities and what learning method best supports their activities within their environment. For example, executive learning focuses on system-level thinking and learning because executives need to view their organizations in a longer-term way (e.g., return on investment), as opposed to viewing learning on an individual, transactional event way. Yet, executives play an integral part in long-term support for technology, as an accelerator. Their role within ROD is to provide the stimulus to support the process of cultural assimilation, and they are also very much a component of strategic integration. Executives do not require as much event-driven reflective change, but they need to be part of the overall "social" structure that paves the way for marrying the benefits of technology with organizational learning. What executives do need to see, are the planned measurable outcomes linked to performance from the investment of coupling organizational learning with technology. The lack of executive involvement and knowledge will be detrimental to the likelihood of making this relationship successful.

Operations, on the other hand, are based more on individual practices of learning. Attempting to incorporate organizational vision and social discourse at this level is problematic until event-driven learning is experienced individually to prove the benefits that can be derived from reflective practices. In addition, there is the problem of the credibility of a learning program. Workers are often wary of new programs designed to enhance their development and productivity. Many question the intentions of the organization and why it is making the investment, especially given what has occurred in corporations over the last 20 years: Layoffs and scandals have riddled organizations and hurt employee confidence in the credibility of employer programs.

Ravell showed us that using reflective practices during events produces accelerated change, driven by technological innovation, which in turn, supports the development of the learning organization. It is important at this level of operations to understand the narrow and pragmatic nature of the way workers think and learn. The way operations personnel are evaluated is also a factor. Indeed, operations personnel are evaluated based on specific performance criteria.

The most complex, yet combined, learning methods relate to the middle management layers. Line managers, within these layers, are engrossed in a double-sided learning infrastructure. On one side, they need to communicate and share with executives what they perceive to be the "overall" issues of the organization. Thus, they need to learn using an organizational learning approach, which is less dependent on event-driven learning and uses reflective practice. Line managers must, along with their senior colleagues, be able to see the business from a more proactive perspective and use social-oriented methods if they hope to influence executives. Details of events are more of an assumed responsibility to them than a preferred way of interacting. In other words, most executives would rather interface with line managers on how they can improve overall operations efficiently and effectively, as opposed to dealing with them on a micro, event-by-event basis. The assumption, then, is that line managers are expected to deal with the details of their operations, unless there are serious problems that require the attention of executives; such problems are usually correlated to failures in the line manager's operations.

On the other side are the daily relationships and responsibilities managers face for their business units. They need to incorporate more individual-based learning techniques that support reflective practices within their operations to assist in the personal development of their staff. The middle management tier described in Figure 5.1 is shown at a summary level and needs to be further described. Figure 5.2 provides a more detailed analysis based on the three types of middle managers described. The figure shows the ratio of organizational learning

Figure 5.2   Organizational/system versus individual learning by middle manager level.

to individual learning based on manager type. The more senior the manager, the more learning is based on systems and social processes.

Knowledge Management

There is an increasing recognition that the competitive advantage of organizations depends on their "ability to create, transfer, utilize, and protect difficult-to-intimate knowledge assets" (Teece, 2001, p. 125). Indeed, according to Bertels and Savage (1998), the dominant logic of the industrial era requires an understanding of how to break the learning barrier to comprehending the information era. While we have developed powerful solutions to change internal processes and organizational structures, most organizations have failed to address the cultural dimensions of the information era. Organizational knowledge creation is a result of organizational learning through strategic processes. Nonaka and Takeuchi (1995) define organizational knowledge as "the capability of a company as a whole to create new knowledge, disseminate it throughout the organization, and embody it in products, services, and systems" (p. 3). Nonaka and Takeuchi use the steps shown in Figure 5.3 to assess the value and chain of events surrounding the valuation of organization knowledge.

Figure 5.3   Nonaka and Takeuchi steps to organizational knowledge.

If we view the Figure 5.3 processes as leading to competitive advantage, we may ask how technology affects the chain of actions that Nonaka and Takeuchi (1995) identify. Without violating the model, we may insert technology and observe the effects it has on each step, as shown in Figure 5.4.

According to Nonaka and Takeuchi (1995), to create new knowledge means to re-create the company, and everyone in it, in an ongoing process that requires personal and organizational self-renewal. That is, knowledge creation is the responsibility of everyone in the organization. The viability of this definition, however, must be questioned. Can organizations create personnel that will adhere to such parameters, and under what conditions will senior management support such an endeavor?

Again, technology has a remarkable role to play in substantiating the need for knowledge management. First, executives are still challenged to understand how they need to deal with emerging technologies as this relates to whether their organizations are capable of using them effectively and efficiently. Knowledge management provides a way for the organization to learn how technology will be used to support innovation and competitive advantage. Second, IT departments need to understand how they can best operate within the larger scope of the organization—they are often searching for a true mission that contains measurable outcomes, as defined by the entire organization, including senior management. Third, both executives and IT staff agree that understanding the uses of technology is a continuous process that should not be utilized solely in a reactionary

---

Knowledge creation: Technology provides more dynamic shifts in knowledge, thus accelerating the number of knowledge-creation events that can occur.

Continuous innovation: Innovations are accelerated because of the dynamic nature of events and the time required to respond—therefore, continuous innovation procedures are more significant to have in each department in order to respond to technological opportunities on an ongoing basis.

Competitive advantage: Technology has generated more global competition. Competitive advantages that depend on technological innovation are more common.

Figure 5.4 Nonaka and Takeuchi organizational knowledge with technology extension.

and event-driven way. Finally, most employees accept the fact that technology is a major component of their lives at work and at home, that technology signifies change, and that participating in knowledge creation is an important role for them.

Again, we can see that technology provides the initiator for understanding how organizational learning is important for competitive advantage. The combination of IT and other organizational departments, when operating within the processes outlined in ROD, can significantly enhance learning and competitive advantage. To expand on this point, I now focus on the literature specifically relating to tacit knowledge and its important role in knowledge management. Scholars theorize knowledge management is an ability to transfer individual tacit knowledge into explicit knowledge. Kulkki and Kosonen (2001) define tacit knowledge as an experience-based type of knowledge and skill and as the individual capacity to give intuitive forms to new things; that is, to anticipate and preconceptualize the future. Technology, by its very definition and form of being, requires this anticipation and preconceptualization. Indeed, it provides the perfect educational opportunity in which to practice the transformation of tacit into explicit knowledge. Tacit knowledge is an asset, and having individual dynamic abilities to work with such knowledge commands a "higher premium when rapid organic growth is enabled by technology" (Teece, 2001, p. 140). Thus, knowledge management is likely to be greater when technological opportunity is richer.

Because evaluating emerging technologies requires the ability to look into the future, it also requires that individuals translate valuable tacit knowledge, and creatively see how these opportunities are to be judged if implemented. Examples of applicable tacit knowledge in this process are here extracted from Kulkki and Kosonen (2001):

- Cultural and social history
- Problem-solving modes
- Orientation to risks and uncertainties
- Worldview organizing principles
- Horizons of expectations

I approach each of these forms of tacit knowledge from the perspective of the components of ROD as shown in Table 5.1.

Table 5.1   Mapping Tacit Knowledge to Responsive Organizational Dynamism

TACIT KNOWLEDGE	STRATEGIC INTEGRATION	CULTURAL ASSIMILATION
Cultural and social history	How the IT department and other departments translate emerging technologies into their existing processes and organization.	
Problem-solving modes	Individual reflective practices that assist in determining how specific technologies can be useful and how they can be applied.	Technology opportunities may require organizational and structural changes to transfer tacit knowledge to explicit knowledge. Utilization of tacit knowledge to evaluate probabilities for success.
Orientation to risks and uncertainties	Technology offers many risks and uncertainties. All new technologies may not be valid for the organization.	Tacit knowledge is a valuable component to fully understand realities, risks, and uncertainties.
Worldviews	Technology has global effects and changes market boundaries that cross business cultures. It requires tacit knowledge to understand existing dispositions on how others work together.	Review how technology affects the dynamics of operations.
Organizing principles	How will new technologies actually be integrated? What are the organizational challenges to "rolling out" products and to implementation timelines? What positions are needed, and who in the organization might be best qualified to fill new responsibilities?	Identify limitations of the organization; that is, tacit knowledge versus explicit knowledge realities.
Horizons of expectations	Individual limitations in the tacit domain that may hinder or support whether a technology can be strategically integrated into the organization.	

It is not my intention to suggest that all technologies should be, or can be, used to generate competitive advantage. To this extent, some technologies may indeed get rejected because they cannot assist the organization in terms of strategic value and competitive advantage. As Teece (2001) states, "Information transfer is not knowledge transfer and information management is not knowledge management, although the former can assist the latter. Individuals and organizations can suffer from information overload" (p. 129). While this is a significant issue for many firms, the ability to have an organization that can select, interpret, and integrate information is a valuable part of knowledge management. Furthermore, advances in IT have propelled much of the excitement surrounding knowledge management. It is important to recognize that learning organizations, reflective practices, and communities of practice all participate in creating new organizational knowledge. This is why knowledge management is so important. Knowledge must be built on its own terms, which requires intensive and laborious interactions among members of the organization.

Change Management

Because technology requires that organizations accelerate their actions, it is necessary to examine how ROD corresponds to theories in organizational change. Burke (2002) states that most organizational change is evolutionary; however, he defines two distinct types of change: planned versus unplanned and revolutionary versus evolutionary. Burke also suggests that the external environmental changes are more rapid today and that most organizations "are playing catch up." Many rapid changes to the external environment can be attributed to emerging technologies, which have accelerated the divide between what an organization does and what it needs to do to remain competitive. This is the situation that creates the need for ROD.

The catching-up process becomes more difficult because the amount of change required is only increasing given ever-newer technologies. Burke (2002) suggests that this catching up will likely require planned and revolutionary change. Such change can be mapped onto much of my work at Ravell. Certainly, change was required; I planned it, and change had to occur. However, the creation of a learning organization, using many of the organizational learning theories addressed in Chapter 4, supports the eventual establishment of an operating organization that can deal with unplanned and evolutionary change. When using technology as the reason for change, it is then important that the components of ROD be integrated with theories of organizational change.

History has shown that most organizational change is not successful in providing its intended outcomes, because of cultural lock-in. *Cultural lock-in* is defined by Foster and Kaplan (2001) as the inability of an organization to change its corporate culture even when there are clear market threats. Based on their definition, then, technology may not be able to change the way an organization behaves, even when there are obvious competitive advantages to doing so. My concern with Foster and Kaplan's conclusion is whether individuals truly understand exactly how their organizations are being affected—or are we to assume that they do understand? In other words, is there a process to ensure that employees understand the impact of not changing? I believe that ROD provides the infrastructure required to resolve this dilemma by establishing the processes that can support ongoing unplanned and evolutionary change.

To best show the relationship of ROD to organizational change theory, I use Burke's (2002) six major points in assisting change in organizations:

1. *Understanding the external environment*: What are competitors and customers' expectations? This is certainly an issue, specifically when tracking whether expected technologies are made available in the client–vendor relationship. But, more critical is the process of how emerging technologies, brought about through external channels, are evaluated and put into production; that is, having a process in place. Strategic integration of ROD is the infrastructure that needs to facilitate the monitoring and management of the external environment.

2. *Evaluation of the inside of the organization*: This directly relates to technology and how it can be best utilized to improve internal operations. While evaluation may also relate to a restructuring of an organization's mission, technology is often an important driver for why a mission needs to be changed (e.g., expanding a market due to e-commerce capabilities).

3. *Readiness of the organization*: The question here is not whether to change but how fast the organization can change to address technological innovations. The ROD arc provides the steps necessary to create organizations that can sustain change as a way of operation, blending strategic integration with cultural assimilation. The maturation of learning: moving toward system-based learning also supports the creation of infrastructures that are vitally prepared for changes from emerging technologies.

4. *Cultural change as inevitable*: Cultural assimilation essentially demands that organizations must dynamically assimilate new technologies and be prepared to evolve their cultures. Such evolution must be accelerated and be systemic within business units, to be able to respond effectively to the rate of change created by technological innovations.

5. *Making the case for change*: It is often difficult to explain why change is inevitable. Much of the need for change can be supported using the reflective practices implemented at Ravell. However, such acceptance is directly related to the process of time. Major events can assist in establishing the many needs for change, as discussed by Burke (2002).

6. *Sustaining change*: Perhaps the strongest part of ROD is its ability to create a process that is evolutionary and systemic. It focuses on driving change to every aspect of the organization and provides organizational learning constructs to address each level of operation. It addresses what Burke (2002) calls the "prelaunch, launch, postlaunch, and sustaining," in the important sequences of organizational change (p. 286).

Another important aspect of change management is leadership. Leadership takes many forms and has multiple definitions. Technology plays an interesting role in how leadership can be presented to organizations, especially in terms of the management style of leadership, or what Eisenhardt and Bourgeois (1988) have coined as "power centralization." Their study examines high-velocity environments in the microcomputer industry during the late 1980s. By *high velocity*, they refer to "those environments in which there is a rapid and discontinuous change in demand, competitors, technology, or regulation, so that information is often inaccurate, unavailable, or obsolete" (p. 738). During the period of their study, the microcomputer industry was undergoing substantial technological change, including the introduction of many new competitors. As it turns out, the concept of high velocity is becoming more the norm today given the way organizations find themselves needing to operate in constant fluxes of velocity. The term *power centralization* is defined as the amount of decision-making control wielded by the CEO. Eisenhardt and Bourgeois's study finds that the more the CEO engages in power-centralized leadership, the greater the degree of politics, which has a negative impact on the strategic performance of the firms examined. This finding suggests that the less democratic the leadership is in high-velocity environments, the less productive the organization will be. Indeed, the study found that when individuals engaged in team learning, political tension was reduced, and the performance of the firms improved.

The structure of ROD provides the means of avoiding the high-velocity problems discovered by the Eisenhardt and Bourgeois (1988) study. This is because ROD allows for the development of more individual learning, as well as system thinking, across the executive ranks of the business. If technology is to continue to establish such high velocities, firms need to examine the Eisenhardt and Bourgeois study for its relevance to everyday operations. They also need to use organizational learning theories as a basis for establishing leadership that can empower employees to operate in an accelerated and unpredictable environment.

Change Management for IT Organizations

While change management theories address a broad population in organizations, there is a need to create a more IT-specific approach to address the unique needs of this group. Lientz and Rea (2004) establish five specific goals for IT change managers:

1. Gain support for change from employees and non-IT managers.
2. Implement change along measurements for the work so that the results of the change are clearly determined.
3. Implement a new culture of collaboration in which employees share more information and work more in teams.
4. Raise the level of awareness of the technology process and work so that there is less of a tendency for reversion.
5. Implement an ongoing measurement process for the work to detect any problems.

Lientz and Rea's (2004) position is that when a new culture is instilled in IT departments, it is particularly important that it should not require massive management intervention. IT people need to be self-motivated to keep up with the myriad accelerated changes in the world of technology. These changes occur inside IT in two critical areas. The first relates to the technology itself. For example, how do IT personnel keep up with new versions of hardware and software? Many times, these changes come in the form of hardware (often called *system*) and software upgrades from vendors who require them to maintain support contracts. The ongoing self-management of how such upgrades and changes will ultimately affect the rest of the organization is a major challenge and one that is difficult to manage top-down. The second area is the impact of new or emerging technologies on business strategy. The challenge is to develop IT personnel who can transform their technical knowledge into business knowledge and, as discussed, take their tacit knowledge and convert it into explicit, strategic knowledge. Further understanding of the key risks to the components of these accelerated changes is provided as follows:

> *System and software version control*: IT personnel must continue to track and upgrade new releases and understand the impact of product enhancements. Some product-related enhancements have no bearing on strategic use; they essentially fix problems in the system or software. On the other hand, some new releases offer new features and functions that need to be communicated to both IT and business managers.
>
> *Existing legacy systems*: Many of these systems cannot support the current needs of the business. This often forces IT staff to figure out how to create what is called "workarounds" (quick fixes) to these systems. This can be problematic given that workarounds might require system changes or modifications to existing software. The risk of these changes, both short and long term, needs to be discussed between user and IT staff communities of practice.
>
> *Software packages (off-the-shelf software)*: Since the 1990s, the use of preprogrammed third-party software packages has become a preferred mode of software use among users. However, many of these packages can be inflexible and do not support the exact processes required by business users. IT personnel need to address users' false expectations about what software packages can and cannot do.

System or software changes: Replacement of systems or software applications is rarely 100% complete. Most often, remnants of old systems will remain. IT personnel can at times be insensitive to the lack of a complete replacement.

Project completion: IT personnel often misevaluate when their involvement is finished. Projects are rarely finished when the software is installed and training completed. IT staff tend to move on to other projects and tasks and lose focus on the likelihood that there will be problems discovered or last-minute requests made by business users.

Technical knowledge: IT staff members need to keep their technical skills up to date. If this is not done, emerging technologies may not be evaluated properly as there may be a lack of technical ability inside the organization to map new technical developments onto strategic advantage.

Pleasing users: While pleasing business users appears to be a good thing, it can also present serious problems with respect to IT projects. What users want, and what they need, may not be the same. IT staff members need to judge when they might need assistance from business and IT management because users may be unfairly requesting things that are not feasible within the constraints of a project. Thus, IT staff must have the ability to articulate what the system can do and what might be advisable. These issues tend to occur when certain business users want new systems to behave like old ones.

Documentation: This, traditionally, is prepared by IT staff and contains jargon that can confuse business users. Furthermore, written procedures prepared by IT staff members do not consider the entire user experience and process.

Training: This is often carried out by IT staff and is restricted to covering system issues, as opposed to the business realities surrounding when, how, and why things are done.

These issues essentially define key risks to the success of implementing technology projects. Much of this book, thus far, has focused on the process of organizational learning from an infrastructure perspective. However, the implementation component of technology possesses new risks to successfully creating an organization that can learn within the needs of ROD. These risks, from the issues enumerated, along with those discussed by Lientz and Rea (2004) are summarized as follows:

Business user involvement: Continuous involvement from business users is necessary. Unfortunately, during the life of a project there are so many human interfaces between IT staff and business users that it is unrealistic to attempt to control these communications through tight management procedures.

Requirements, definition, and scope: These relate to the process by which IT personnel work with business users to determine exactly what software and systems need to accomplish. Determining requirements is a process, not a predetermined list that business users will necessarily have available to them. The discourse that occurs in conversations is critical to whether such communities are capable of developing requirements that are unambiguous in terms of expected outcomes.

Business rules: These rules have a great effect on how the organization handles data and transactions. The difference between requirements and business rules is subtle. Specifically, business rules, unlike requirements, are not necessarily related to processes or events of the business. As such, the determination of business rules cannot be made by reviewing procedures; for example, all account numbers must be numeric.

Documentation and training materials: IT staff members need to interact with business users and establish joint processes that foster the development of documentation and training that best fit user needs and business processes.

Data conversion: New systems and applications require that data from legacy systems be converted into the new formats. This process is called *data mapping;* IT staff and key business users review each data field to ensure that the proper data are represented correctly in the new system. IT staff members should not be doing this process without user involvement.

Process measurement: Organizations typically perform a post-completion review after the system or software application is installed. Unfortunately, this process measurement should occur during and after project completion.

IT change management poses some unique challenges to implementing organizational learning, mostly because managers cannot conceivably be available for all of the risks identified. Furthermore, the very nature of new technologies requires that IT staff members develop the ability to self-manage more of their daily functions and interactions, particularly with other staff members outside the IT department. The need for self-development is even more critical because of the existence of technological dynamism, which focuses on dynamic and unpredictable transactions that often must be handled directly by IT staff members and not their managers. Finally, because so many risks during technology projects require business user interfaces, non-IT staff members also need to develop better and more efficient self-management than they are accustomed to doing. Technological dynamism, then, has established another need for change management theory. This need relates to the implementation of self-development methods. Indeed, part of the reason for the lack of success of IT projects can be attributed to the inability of the core IT and business staff to perform in a more dynamic way. Historically, more management cannot provide the necessary learning and reduction of risk.

The idea of self-development became popular in the early 1980s as an approach to the training and education of managers, and managers to be. Thus, the focus of management self-development is to increase the ability and willingness of managers to take responsibility for themselves, particularly for their own learning (Pedler et al., 1988). I believe that management self-development theory can be applied to nonmanagers, or to staff members, who need to practice self-management skills that can assist them in transitioning to operating under the conditions of technological dynamism.

Management self-development draws on the idea that many people emphasize the need for learner centeredness. This is an important concept in that it ties self-development theory to organizational learning, particularly to the work of Chris Argyris and Malcolm Knowles. The concept of learner centeredness holds that individuals must take prime responsibility for their own learning: when and how to learn. The teacher (or manager) is assigned the task of facilitator—a role that fosters guidance as opposed to direct initiation of learning. In many ways, a facilitator can be seen as a mentor whose role it is to guide an individual through various levels of learning and individual development.

What makes self-development techniques so attractive is that learners work on actual tasks and then reflect on their own efforts. The methods of reflective practice theory, therefore, are applicable and can be integrated with self-development practices. Although self-development places the focus on the individual's own efforts, managers still have responsibilities to mentor, coach, and counsel their staff. This support network allows staff to receive appropriate feedback and guidance. In many ways, self-development relates to the professional process of apprenticeship but differs from it in that the worker may not aspire to become the manager but may wish simply to develop better management skills. Workers are expected to make mistakes and to be guided through a process that helps them reflect and improve. This is why self-development can be seen as a management issue as opposed to just a learning theory.

A mentor or coach can be a supervisor, line manager, director, or an outside consultant. The bottom line is that technological dynamism requires staff members who can provide self-management to cope with constant project changes and risks. These individuals must be able to learn, be self-aware of what they do not know, and possess enough confidence to initiate the required learning and assistance that they need to be successful (Pedler et al., 1988). Self-development methods, like other techniques, have risks. Most notable, is the initial decrement in performance followed by a slow increment as workers become more comfortable with the process and learn from their mistakes. However, staff members must be given support and time to allow this process to occur; self-development is a trial-and-error method founded on the basis of mastery learning (i.e., learning from one's mistakes). Thus, the notion of self-development is both continuous and discontinuous and must be implemented in a series of phases, each having unique outcomes and maturity. The concept of self-development is also consistent with the ROD arc, in which early phases of maturation require more individual learning, particularly reflective practices. Self-development, in effect, becomes a method of indirect management to assist in personal transformation. This personal transformation will inevitably better prepare individuals to participate in group- and organizational-level learning at later stages of maturation.

The first phase of establishing a self-development program is to create a "learning-to-learn" process. Teaching individuals to learn is a fundamental need before implementing self-development techniques. Mumford (1988) defines learning to learn as

1. Helping staff to understand the stages of the learning process and the pitfalls to not learning
2. Helping staff to find their own preferences to learning
3. Assisting staff in understanding their present learning preferences and how to deal with, and overcome, learning weaknesses
4. Helping staff to build on their learning experience and apply it to their current challenges in their job

The first phase of self-development clearly embraces the Kolb (1999) Learning Style Inventory and the applied individual learning wheel that were introduced in Chapter 4. Thus, all staff members should be provided with both of these learning wheels, made aware of their natural learning strengths and weaknesses, and provided with exercises to help them overcome their limitations. Most important is that the Kolb system will make staff aware of their shortfalls with learning. The applied individual learning wheel will provide a perspective on how individuals can link generic learning preferences into organizational learning needs to support ROD.

The second phase of self-development is to establish a formal learning program in which staff members

1. Are responsible for their own learning, coordinated with a mentor or coach
2. Have the right to determine how they will meet their own learning needs, within available resources, time frames, and set outcomes
3. Are responsible for evaluating and assessing their progress with their learning

In parallel, staff coaches or mentors

1. Have the responsibility to frame the learning objectives so that they are consistent with agreed-on individual weaknesses

2. Are responsible for providing access and support for staff
3. Must determine the extent of their involvement with mentoring and their commitment to assisting staff members achieve stated outcomes
4. Are ultimately responsible for the evaluation of individual's progress and success

This program must also have a formal process and structure. According to Mossman and Stewart (1988), formal programs, called self-managed learning (SML), need the following organization and materials:

1. Staff members should work in groups as opposed to on their own. This is a good opportunity to intermix IT and non-IT staff with similar issues and objectives. The size of these groups is (typically) from four to six members. Groups should meet every two–three weeks, and should develop what are known as *learning contracts*. Learning contracts specifically state what the individual and management have agreed on. Essentially, the structure of self-development allows staff members to experience communities of practice, which by their very nature, will also introduce them to group learning and system-level thinking.

2. Mentors or coaches should preside over a group as opposed to presiding over just one individual. There are two benefits to doing this: (1) There are simply economies of scale for which managers cannot cover staff on an individual basis, and (2) facilitating a group with similar objectives benefits interaction among the members. Coaches obviously need to play an important role in defining the structure of the sessions, in offering ideas about how to begin the self-development process, and in providing general support.

3. Staff members need to have workbooks, films, courses, study guides, books, and specialists in the organization, all of which learners can use to help them accomplish their goals.

4. Typically, learning contracts will state the assessment methods. However, assessment should not be limited only to individuals but also should include group accomplishments.

An SML should be designed to ensure that the learning program for staff members represents a commitment by management to a formal process, that can assist in the improvement of the project teams.

The third phase of self-development is evaluation. This process is a mixture of individual and group assessments from phase II, coupled with assessments from actual practice results. These are results from proven outcomes during normal workday operations. To garner the appropriate practice evaluation, mentors and coaches must be involved in monitoring results and noting the progress on specific events that occur. For example, if a new version of software is implemented, we will want to know if IT staff and business users worked together to determine how and when it should be implemented. These results need to be formally communicated back to the learning groups. This process needs to be continued on an ongoing basis to sustain the effects of change management. Figure 5.5 represents the flow of the three phases of the process.

The process for self-development provides an important approach in assisting staff to perform better under the conditions of technological dynamism. It is one thing to teach reflective practice; it is another

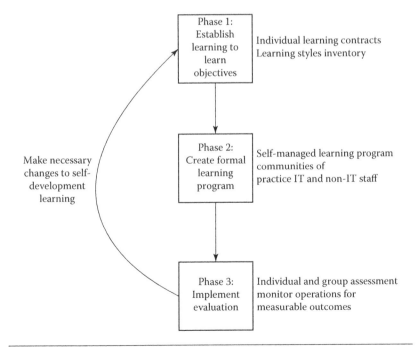

Figure 5.5 Phases of self-development.

to get staff members to learn how to think in a manner that takes into consideration the many risks that have plagued systems and software projects for decades. While the role of management continues to play a major part in getting things done within strategic objectives, self-development can provide a strong learning method, that can foster sustained bottom-up management, which is missing in most learning organizations.

The Ravell case study provides some concrete evidence on how self-development techniques can indeed get results. Because of the time pressures at Ravell, I was not able to invest in the learning-to-learn component at the start of the process. However, I used informal methods to determine the learning preferences of the staff. This can be accomplished through interviews in which staff responses can provide a qualitative basis for evaluating how specific personnel prefer to learn. This helped me to formulate a specific training program that involved group meetings with IT and non-IT-oriented groups.

In effect, phase II at Ravell had two communities. The first community was the IT staff. We met each week to review progress and to set short-term objectives of what the community of IT wanted to accomplish. I acted as a facilitator, and although I was in a power position as their manager, I did not use my position unless there were clear signs of resistance in the team (which there were in specific situations). The second community was formed with various line manager departments. This is where I formed "dotted-line" reporting structures, which required IT staff members also to join other communities of practice. This proved to be an invaluable strategy because it brought IT and business users together and formed the links that eventually allowed IT staff members to begin to learn and to form relationships with the user community, which fostered reflective thinking and transformation.

As stated, there are setbacks at the start of any self-development program, and the experience at Ravell was no exception. Initially, IT staff members had difficulty understanding what was expected of them; they did not immediately perceive the learning program as an opportunity for their professional growth. It was through ongoing, motivated discourse in and outside of the IT community that helped achieve measurable increments of self-developmental growth. Furthermore, I found it necessary to integrate individual coaching sessions with IT staff. While group sessions were useful, they were not a substitute for individual discussions, which at times allowed IT staff members to personally discuss their concerns and learning requirements. I found the process to be ultimately valuable, and I maintained the role of coach, as opposed to that of a manager who tells IT staff members what to do in every instance. I knew that direct management only would never allow for the development of learning.

Eventually, self-development through discourse will foster identity development. Such was the case at Ravell, where both user and IT groups eventually came together to form specific and interactive communities of practice. This helped form a clearer identity for IT staff members, and they began to develop the ability to address the many project risk issues that I defined in this chapter. Most important for the organization was that Ravell phase I built the foundation for later phases that required more group and system thinking among the IT ranks.

Evaluation of the performance at Ravell (phase III of the self-development process) was actually easier than expected, which means that if the first two phases are successful, evaluation will naturally be easy to determine. As reflective thinking became more evident in the group, it was easier to see the growth in transformative behavior; the IT groups became more proactive and critical by themselves, without necessarily needing my input. In fact, my participation fell into more of a supporter role; I was asked to participate more when I felt needed to provide a specific task for the group. Evaluation based on performance was also easier to determine, mainly because we had formed interdepartmental communities and because of the relationships I established with line managers.

Another important decision we made and one that nurtured our evaluation capabilities was the fact that line managers often joined our IT staff meetings. So, getting feedback on actual results was always open for discussion.

Viewing self-development in the scope of organizational learning and management techniques provides an important support method for later development in system thinking. The Ravel experience did just that, as the self-development process inevitably laid the foundation for more sophisticated organizational learning, required as a business matures under ROD.

Social Networks and Information Technology

The expansion of social networks, through the use of technological innovations, has substantially changed the way information flows in and out of a business community. Some companies, particularly in the financial services communities, have attempted to "lock out" social network capabilities. These attempts are ways for organizations to control, as opposed to change, behavior. Historically, such controls to enforce compliance have not worked. This is particularly relevant because of the emergence of a younger generation of workers who use social networking tools as a regular way to communicate and carry out discourse. Indeed, social networking has become the main vehicle for social discourse both inside and outside organizations. There are those who feel that the end of confidentiality may be on the horizon. This is not to suggest that technology executives give up on security—we all know this would be ludicrous. On the other hand, the increasing pressure to "open" the Web will inevitably become too significant to ignore. Thus, the technology executive of the future must be prepared to provide desired social and professional networks to their employees while figuring out how to minimize risk—certainly not an easy objective. Organizations will need to provide the necessary learning techniques to help employees understand the limits of what can be done.

We must remember that organizations, governments, and businesses have never been successful at controlling the flow of information to any population to or from any specific interest group—inevitably, information flows through. As stated by Cross and Thomas (2009), "The network perspective could trigger new approaches to organization design at a time when environmental and competitive conditions seem to be exhausting conventional wisdom" (p. 186). Most important is the understanding that multinational organizations need to think globally and nationally at the same time. To do this, employees must transform their behavior and how they interact. Controlling access does not address this concern; it only makes communication more difficult and therefore does not provide a solution. Controls typically manifest themselves in the form of new processes and procedures. I often see technology executives proclaiming the need to change processes in the name of security without really understanding that they are not providing a solution, but rather, fostering new procedures that will allow individuals to evade the new security measures. As Cross and Thomas (2009) point out, "Formal structures often overlook the fact that every formal organization has in its shadow an informal or 'invisible' organization" (p. 1). Instead, technology executives concerned with security, need to focus on new organizational design to assist businesses to be "social network ready." ROD must then be extended to allow for the expansion of social network integration, including, but not limited to, such products as LinkedIn, Facebook, and Twitter. It may also be necessary to create new internal network infrastructures that specifically cater to social network communication.

Many software application companies have learned that compatibility in an open systems environment is a key factor for successful deployment of an enterprise-wide application solution. Thus, all applications developed within or for an organization need to have compatibility with the common and popular social network products. This popularity is not static, but rather, a constant process of determining which products will become important social networks that the company may want to leverage. We see social networks having such an impact within the consumer environment—or what we can consider to be the "market." I explained in my definition of ROD that it is the acceleration of market changes—or the changing relationship between a buyer and seller—that dictates the successes and failures of businesses. That said, technology executives must focus their attention on how such networks will require their organizations to embrace them. Obviously, this change carries risks. Adapting too early could be overreacting to market hype, while lagging could mean late entry.

The challenge, then, for today's technology leaders is to create dynamic, yet functional, social networks that allow businesses to compete while maintaining the controls they must have to protect themselves. The IT organization must concentrate on how to provide the infrastructure that allows these dynamic connections to be made without overcontrol. The first mission for the technology executive is to negotiate this challenge by working with the senior management of the organization to reach consensus on the risk factors. The issues typically involve the processes, behavior patterns, and risks shown in Figure 5.6.

Ultimately, the technology executive must provide a new road map that promotes interagency and cross-customer collaboration in a way

Business process	Aspired behavior patterns	Risks
Design a social network that allows participants to respond dynamically to customer and business needs	Users understand the inherent limits to what can be communicated outside the organization, limit personal transactions, and use judgment when foreign e-mails are forwarded.	Users cannot properly determine the ethics of behavior and will not take the necessary precautions to avoid exposing the organization to outside security breaches.
Discern which critical functions are required for the social network to work effectively and maintain the firm's competitive positioning	Users are active and form strategic groups (communities of practice) that define needs on a regular basis and work closely with IT and senior management.	Users cannot keep up with changes in social networks, and it is impossible to track individual needs and behaviors.
Provide a network design that can be scaled as needs change within the budget limitations of the organization	The organization must understand that hard budgets for social networking may not be feasible. Rather, the network needs are dynamic, and costs must be assessed dynamically within the appropriate operating teams in the organizations.	Reality tells us that all organizations operate within budget limitations. Large organizations find it difficult to govern dynamically, and smaller organizations cannot afford the personnel necessary to manage dynamically.
Create a social network that "flattens" the organization so that all levels are accessible	Particularly large organizations need to have a network that allows its people better access to its departments, talent, and management. In the 1980s, the book *In Search of Excellence* (Peters & Waterman, 1982) was the first effort to present the value of a "flatter" organizational structure. Social networks provide the infrastructure to make this a reality.	With access come the challenges of responding to all that connect to the system. The organization needs to provide the correct etiquette of how individuals respond dynamically without creating anarchy.

Figure 5.6   Social network management issues.

that will assist the organization to attain a ROD culture. Social networks are here to stay and will continue to necessitate 24/7 access for everyone. This inevitably raises salient issues relating to the management structure within businesses and how best to manage them.

In Chapter 2, I defined the IT dilemma in a number of contexts. During an interview, a chief executive raised an interesting issue that relates to the subject: "My direct reports have been complaining that because of all this technology that they cannot get away from—that their days never seem to end." I responded to this CEO by asking,

"Why are they e-mailing and calling you? Is it possible that technology has exposed a problem that has always existed?" The CEO seemed surprised at my response and said, "What do you mean?" Again, I responded by suggesting that technology allowed access, but perhaps, that was not really the problem. In my opinion, the real problem was a weakness in management or organizational structure. I argued that good managers build organizations that should handle the questions that were the subject of these executives' complaints. Perhaps the real problem was that the organization or management was not handling day-to-day issues. This case supports my thesis that technology dynamism requires reevaluation of how the organization operates and stresses the need to understand the cultural assimilation abilities of dealing with change.

Another interesting aspect of social networks is the emergence of otherwise invisible participants. Technology-driven networks have allowed individuals to emerge not only because of the access determinant but also because of statistics. Let me be specific. Network traffic can easily be tracked, as can individual access. Even with limited history, organizations are discovering the valued members of their companies simply by seeing who is active and why. This should not suggest that social networks are spy networks. Indeed, organizations need to provide learning techniques to guide how access is tracked and to highlight the value that it brings to a business. As with other issues, the technology executive must align with other units and individuals; the following are some examples:

- *Human resources (HR)*: This department has specific needs that can align effectively with the entire social network. Obviously, there are compliance issues that limit what can be done over a network. Unfortunately, this is an area that requires reassessment: In general, governance and controls do not drive an organization to adopt ROD. There are other factors related to the HR function. First, is the assimilation of new employees and the new talents that they might bring to the network. Second, is the challenge of adapting to ongoing change within the network. Third, is the knowledge lost of those who leave the organization yet may still want to participate socially within the organization (friends of the company).

- *Gender*: Face-to-face meetings have always shown differences in participation by gender. Men tend to dominate meetings and the positions they hold in an organization. However, the advent of social virtual networks has begun to show a shift in the ways women participate and hold leadership positions among their peers. In an article in *Business Week* (May 19, 2008), Auren Hoffman reports that women dominate social network traffic. This may result in seeing more women-centric communication. The question, then, is whether the expansion of social networks will give rise to more women in senior management positions.

- *Marketing*: The phenomenon of social networking has allowed for the creation of more targeted connectivity; that is, the ability to connect with specific clients in special ways. Marketing departments are undergoing an extraordinary transformation in the way they target and connect with prospective customers. The technology executive is essentially at the center of designing networks that provide customizable responses and facilitate complex matrix structures. Having such abilities could be the differentiator between success and failure for many organizations.

One can see that the expansion of social networks is likely to have both good and bad effects. Thus far, in this section I have discussed the good. The bad relates to the expansion of what seems to be an unlimited network. How does one manage such expansion? The answer lies within the concept of alignment. Alignment has always been critical to attain organizational effectiveness. The heart of alignment is dealing with cultural values, goals, and processes that are key to meet strategic objectives (Cross & Thomas, 2009). While the social network acts to expose these issues, it does not necessarily offer solutions to these differences. Thus, the challenge for the technology executive of today is to balance the power of social networks while providing direction on how to deal with alignment and control—not an easy task but clearly an opportunity for leadership. The following chapters offer some methods to address the challenges discussed in this chapter, and the opportunities they provide for technology executives.

6

ORGANIZATIONAL TRANSFORMATION AND THE BALANCED SCORECARD

Introduction

The purpose of this chapter is to examine the nature of organizational transformation, how it occurs, and how it can be measured. Aldrich (2001) defines organizational transformation along three possible dimensions: changes in goals, boundaries, and activities. According to Aldrich, transformations "must involve a qualitative break with routines and a shift to new kinds of competencies that challenge existing organizational knowledge" (p. 163). He warns us that many changes in organizations disguise themselves as transformative but are not. Thus, focusing on the qualifications of authentic or substantial transformation is key to understanding whether it has truly occurred in an organization. Technology, as with any independent variable, may or may not have the capacity to instigate organizational transformation. Therefore, it is important to integrate transformation theory with responsive organizational dynamism (ROD). In this way, the measurable outcomes of organizational learning and technology can be assessed in organizations that implement ROD. Most important in this regard, is that organizational transformation, along with knowledge creation, be directly correlated to the results of implementing organizational learning. That is, the results of using organizational learning techniques must result in organizational transformation.

Organizational transformation is significant for three key reasons:

1. Organizations that cannot change will fundamentally be at risk against competitors, especially in a quickly changing market.

2. If the organization cannot evolve, it will persist in its norms and be unwilling to change unless forced to do so.
3. If the community population is forced to change and is constrained in its evolutionary path, it is likely that it will not be able to transform and thus, will need to be replaced.

Aldrich (2001) establishes three dimensions of organizational transformation. By examining them, we can apply technology-specific changes and determine within each dimension what constitutes authentic organizational transformation.

1. *Goals*: There are two types of goal-related transformations: (a) change in the market or target population of the organization; (b) the overall goal of the organization itself changes. I have already observed that technology can affect the mission of an organization, often because it establishes new market niches (or changes them). Changed mission statements also inevitably modify goals and objectives.
2. *Boundaries*: Organizational boundaries transform when there is expansion or contraction. Technology has historically expanded domains by opening up new markets that could not otherwise be reached without technological innovation. E-business is an example of a transformation brought about by an emerging technology. Of course, business can contract as a result of not assimilating a technology; technology also can create organizational transformation.
3. *Activity systems*: Activity systems define the way things are done. They include the processing culture, such as behavioral roles. Changes in roles and responsibilities alone do not necessarily represent organizational transformation unless it is accompanied by cultural shifts in behavior. The cultural assimilation component of ROD provides a method with which to facilitate transformations that are unpredictable yet evolutionary. Sometimes, transformations in activity systems deriving from technological innovations can be categorized by the depth and breadth of its impact on other units. For example, a decision could be made to use technology as part of a total quality management (TQM)

effort. Thus, activity transformations can be indirect and need to be evaluated based on multiple and simultaneous events.

Aldrich's (2001) concept of organizational transformation bears on the issue of frequency of change. In general, he concludes that the changes that follow a regular cycle are part of normal evolution and "flow of organizational life" (p. 169) and should not be treated as transformations. Technology, on the other hand, presents an interesting case in that it can be perceived as normal in its persistence and regularity of change while being unpredictable in its dynamism. However, Aldrich's definition of transformation poses an interesting issue for determining transformations resulting from technological innovations. Specifically, under what conditions is a technological innovation considered to have a transformative effect on the organization? And, when is it to be considered as part of regular change? I refer to Figure 6.1, first presented in Chapter 3 on driver and supporter life cycles to respond to this question.

The flows in this cycle can be used as the method to determine technological events that are normal change agents versus transformative ones. To understand this point, one should view all driver-related technologies as transformational agents because they, by definition, affect strategic innovation and are approved based on return on investment (ROI). Aldrich's (2001) "normal ebb and flows" represent the "mini-loops" that are new enhancements or subtechnologies, which are part of normal everyday changes necessary to mature a

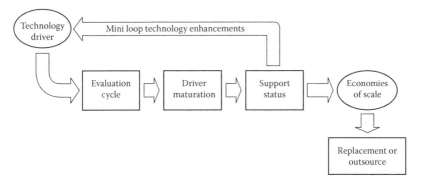

Figure 6.1 Driver-to-supporter life cycle.

technological innovation. Thus, driver variables that result from mini-loops, would not be considered transformational agents of change.

It is important to recognize that Aldrich's (2001) definition of organizational transformation should not be confused with theories of transformative learning. As West (1996) proclaims, "The goal of organizational learning is to transform the organization" (p. 54). The study of transformative learning has been relevant to adult education, and has focused on individual, as opposed to organizational, development and learning. Thus, transformative learning has been better integrated in individual learning and reflective practice theories than in organizational ones. While these modes of learning are related to the overall learning in organizations, they should not be confused with organizations that are attempting to realize their performance objectives.

Yorks and Marsick (2000) offer two strategies that can produce transformative learning for individuals, groups, or organizations: action learning and collaborative inquiry. I covered action science in Chapter 4, particularly reflective practices, as key interventions to foster both individual and group evolution of learning, specifically in reference to how to manage ROD. Aspects of collaborative inquiry are applied to later stages of maturation and to more senior levels of management based on systems-level learning. As Yorks and Marsick (2000) state, "For the most part the political dimensions of how the organization functions is off limits, as are discussions of larger social consequences" (p. 274).

Technological innovations provide acceleration factors and foster the need for ROD. Technology also furnishes the potential tangible and measurable outcomes necessary to normalize York and Marsick's (2000) framework for transformative learning theory into organizational contexts as follows:

1. Technology, specifically e-business, has created a critical need for organizations to engage with clients and individuals in a new interactive context. This kind of discourse has established accelerated needs, such as understanding the magnitude of alternative courses of action between customer and vendor. The building of sophisticated intranets (internal Internets) and their evolution to assimilate with other Internet operations

has also fueled the need for learning to occur more often than before and at organizational level.

Because technology can produce measurable outcomes, individuals are faced with accelerated reflections about the cultural impact of their own behaviors. This is directly related to the implementation of the cultural assimilation component of ROD, by which individuals determine how their behaviors are affected by emerging technologies.

2. Early in the process of implementing strategic integration, reflective practices are critical for event-driven technology projects. These practices force individuals to continually reexamine their existing meaning perspectives (specifically, their views and habits of mind). Individual reflection in, on, and to practice will evolve to system-level group and organizational learning contexts, as shown in the ROD arc.

3. The process of moving from individual to system-level learning during technology maturation is strengthened by the learners' abilities to comprehend why historical events have influenced their existing habits of mind.

4. The combination of strategic integration and cultural assimilation lays the foundation for organizational transformation to occur. Technology provides an appropriate blend of being both strategic and organizational in nature, thus allowing learners to confront their prior actions and develop new practices.

Aldrich (2001) also provides an interesting set of explanations for why it is necessary to recognize the evolutionary aspect of organizational transformations. I have extended them to operate within the context of ROD, as follows:

Variation: Defined as "change from current routines and competencies and change in organizational forms" (Aldrich, 2001, p. 22). Technology provides perhaps the greatest amount of variation in routines and thereby establishes the need for something to manage it: ROD. The higher the frequency of variation, the greater the chance that organizational transformation can occur. Variation is directly correlated to cultural assimilation.

Selection: This is the process of determining whether to use a technology variation. Selections can be affected by external (outside the organization) and internal (inside the organization) factors, such as changes in market segments or new business missions, respectively. The process of selection can be related to the strategic integration component of ROD.

Retention: Selected variations are retained or preserved by the organization. Retention is a key way of validating whether organizational transformation has occurred. As Aldrich states: "Transformations are completed when knowledge required for reproducing the new form is embodied in a community of practice" (p. 171).

Because of the importance of knowledge creation as the basis of transformation, communities of practice are the fundamental structures of organizational learning to support organizational transformation. Aldrich (2001) also goes beyond learning; he includes policies, programs, and networks as parts of the organizational transformative process. Figure 6.2 shows Aldrich's evolutionary process and its relationship to ROD components.

Thus, we see from Figure 6.2 the relationships between the processes of creating organizational transformation, the stages required to reach it, the ROD components in each stage, and the corresponding organizational learning method that is needed. Notice that the mapping of organizational learning methods onto Aldrich's (2001) scheme for organizational transformation can be related to the ROD arc. It shows us that as we get closer to retention, organizational learning evolves from an individual technique to a system/organizational learning perspective. Aldrich's model is consistent with my driver-versus-supporter concept. He notes, "When the new form becomes a taken-for-granted aspect of every day life in the organization, its legitimacy is assumed" (p. 175).

Hence, the assimilation of new technologies cannot be considered transformative until it behaves as a supporter. Only then can we determine that the technology has changed organizational biases and norms. Representing the driver and supporter life cycle to include this important relationship is shown in Figure 6.3.

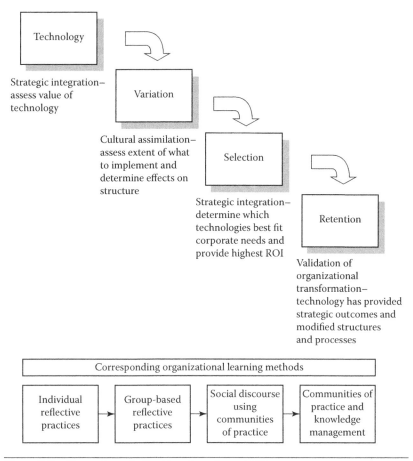

Figure 6.2   Stages of organizational transformation and ROD.

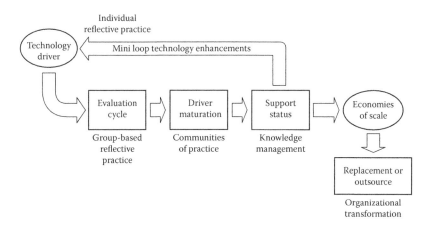

Figure 6.3   Organizational transformation in the driver-to-supporter life cycle.

Methods of Ongoing Evaluation

If we define organizational transformation as the retention of knowledge within the body of communities of practice, the question to be answered is how this retention actually is determined in practice. The possibility often occurs that transformations are partial or in some phase of completion. This would mean that the transformation is incomplete or needs to continue along some phase of approach. Indeed, cultural assimilation does not occur immediately, but rather, over periods of transition. Much of the literature on organizational transformation does not address the practical aspects of evaluation from this perspective. This lack of information is particularly problematic with respect to technology, since so much of how technology is implemented relates to phased steps that rarely happen in one major event. Thus, it is important to have some method of ongoing evaluation to determine the extent of transformation that has occurred and which organizational learning methods need to be applied to help continue the process toward complete transformation.

Aldrich's (2001) retention can also be misleading. We know that organizational transformation is an ongoing process, especially as advocated in ROD. It is probable that transformations continue and move from one aspect of importance to another, so a completed transformation may never exist. Another way of viewing this concept is to treat transformations as event milestones. Individuals and communities of practice are able to track where they are in the learning process. It also fits into the phased approach of technology implementation. Furthermore, the notion of phases allows for integration of organizational transformation concepts with stage and development theories. With the acceptance of this concept, there needs to be a method or model that can help organizations define and track such phases of transformation. Such a model would also allow for mapping outcomes onto targeted business strategies. Another way of understanding the importance of validating organizational transformation is to recognize its uniqueness, since most companies fail to execute their strategies.

The method that can be applied to the validation of organizational transformation is a management tool called the balanced scorecard. The balanced scorecard was introduced by Kaplan and Norton (2001) in the early 1990s as a tool to solve measurement problems. The ability of an organization to develop and operationalize its intangible assets has become more and more a critical component for success. As I have already expressed regarding the work of Lucas (1999), financial measurement may not be capable of capturing all IT value. This is particularly true in knowledge-based theories. The balanced score-card can be used as a solution for measuring outcomes that are not always financial and tangible. Furthermore, the balanced scorecard is a "living" document that can be modified as certain objectives or measurements require change. This is a critical advantage because, as I have demonstrated, technology projects often change in scope and in objectives as a result of internal and external factions.

The ultimate value, then, of the balanced scorecard, in this context, is to provide a means for evaluating transformation not only for measuring completion against set targets but also for defining how expected transformations map onto the strategic objectives of the organization. In effect, it is the ability of the organization to execute its strategy. Before explaining the details of how a balanced scorecard can be applied specifically to ROD, I offer Figure 6.4, which shows

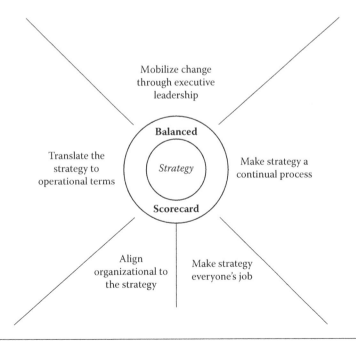

Figure 6.4 Balanced scorecard. (From Kaplan, R.S., & Norton, D.P., *The Strategy-Focused Organization*, Harvard University Press, Cambridge, MA, 2001.)

exactly where the scorecard fits into the overall picture of transition-ing emerging technologies into concrete strategic benefit.

The generic objectives of a balanced scorecard are designed to cre-ate a strategy-focused organization. Thus, all of the objectives and measurements should be derived from the vision and strategy of the organization (Kaplan & Norton, 2001). These measurements are based on the fundamental principles of any strategically focused orga-nization and on alignment and focus. Kaplan and Norton define these principles as the core of the balanced scorecard:

1. *Translate the strategy to operational terms*: This principle includes two major components that allow an organization to define its strategy from a cause-and-effect perspective using a strategy map and scorecard. Thus, the strategy map and its corresponding balanced scorecard provide the basic measure-ment system.

2. *Align the organization to the strategy*: Kaplan and Norton define this principle as favoring synergies among organiza-tional departments that allow communities of practice to have a shared view, and common understanding of their roles.

3. *Make strategy everyone's everyday job*: This principle supports the notion of a learning organization that requires everyone's participation, from the chief executive officer (CEO) to cleri-cal levels. To accomplish this mission, the members of the organization must be aware of business strategy; individuals may need "personal" scorecards and a matching reward sys-tem for accomplishing the strategy.

4. *Make strategy a continual process*: This process requires the linking of important, yet fundamental, components, includ-ing organizational learning, budgeting, management reviews, and a process of adaptation. Much of this principle falls into the areas of learning organization theories that link learning and strategy in ongoing perpetual cycles.

5. *Mobilize change through executive leadership*: This principle stresses the need for a strategy-focused organization that incorporates the involvement of senior management and can mobilize the organization and provide sponsorship to the overall process.

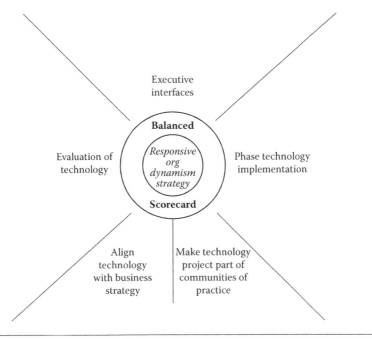

Figure 6.5 Balanced scorecard ROD.

Using the core balanced scorecard schematic, I have modified it to operate with technology and ROD, as shown in Figure 6.5.

1. *Evaluation of technology*: The first step is to have an infrastructure that can determine how technology fits into a specific strategy. Once this is targeted, the evaluation team needs to define it in operational terms. This principle requires the strategic integration component of ROD.

2. *Align technology with business strategy*: Once technology is evaluated, it must be integrated into the business strategy. This involves ascertaining whether the addition of technology will change the current business strategy. This principle is also connected to the strategic integration component of ROD.

3. *Make technology projects part of communities of practice*: Affected communities need to be strategically aware of the project. Organizational structures must determine how they distribute rewards and objectives across departments. This principle requires the cultural assimilation component of ROD.

4. *Phased-in technology implementation*: Short- and long-term project objectives are based on driver and supporter life cycles.

This will allow organizational transformation phases to be linked to implementation milestones. This principle maps onto the cultural assimilation component of ROD.

5. *Executive interface*: CEO and senior managers act as executive sponsors and project champions. Communities of practice and their common "threads" need to be defined, including middle management and operations personnel, so that top-down, middle-up-down, and bottom-up information flows can occur.

The balanced scorecard ultimately provides a framework to view strategy from four different measures:

1. *Financial*: ROI and risk continue to be important components of strategic evaluation.
2. *Customer*: This involves the strategic part of how to create value for the customers of the organization.
3. *Internal business processes*: This relates to the business processes that provide both customer satisfaction and operational efficiency.
4. *Learning and growth*: This encompasses the priorities and infrastructure to support organizational transformation through ROD.

The generic balanced scorecard framework needs to be extended to address technology and ROD. I propose the following adjustments:

1. *Financial*: Requires the inclusion of indirect benefits from technology, particularly as Lucas (1999) specifies, in nonmonetary methods of evaluating ROI. Risk must also be factored in, based on specific issues for each technology project.
2. *Customer*: Technology-based products are integrated with customer needs and provide direct customer package interfaces. Further, web systems that use the Internet are dependent on consumer use. As such, technology can modify organizational strategy because of its direct effect on the customer interface.
3. *Internal business processes*: Technology requires business process reengineering (BPR), which is the process of reevaluating existing internal norms and behaviors before designing a

new system. This new evaluation process addresses customers, operational efficiencies, and cost.

4. *Learning and growth*: Organizational learning techniques, under the umbrella of ROD, need to be applied on an ongoing and evolutionary basis. Progress needs to be linked to the ROD arc.

The major portion of the balanced scorecard strategy is in its initial design; that is, in translating the strategy or, as in the ROD scorecard, the evaluation of technology. During this phase, a strategy map and actual balanced scorecards are created. This process should begin by designing a balanced scorecard that articulates the business strategy. Remember, every organization needs to build a strategy that is unique and based on its evaluation of the external and internal situation (Olve et al., 2003). To clarify the definition of this strategy, it is easier to consider drawing the scorecard initially in the form of a strategy map. A generic strategy map essentially defines the components of each perspective, showing specific strategies within each one, as shown in Figure 6.6.

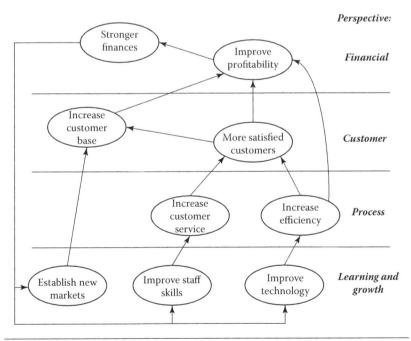

Figure 6.6 Strategy map. (From Olve, N., et al., *Making Scorecards Actionable: Balancing Strategy and Control*, Wiley, New York, 2003.)

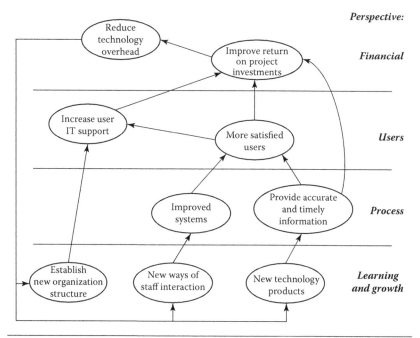

Figure 6.7   Technology strategy map.

We can apply the generic strategy map to an actual case study, Ravell phase I, as shown in Figure 6.7.

Recall that Ravell phase I created a learning organization using reflective practices and action science. Much of the organization transformation at Ravell was accelerated by a major event—the relocation of the company. The move was part of a strategic decision for the organization, specifically the economies of scale for rental expense and an opportunity to retire old computers and replace them with a much needed state-of-the-art network. Furthermore, there was a grave need to replace old legacy applications that were incapable of operating on the new equipment and were also not providing the competitive advantage that the company sought. In using the strategy map, a balanced scorecard can be developed containing the specific outcomes to achieve the overall mission. The balanced scorecard is shown in Figure 6.8.

The Ravell balanced scorecard has an additional column that defines the expected organizational transformation from ROD. This model addresses the issue of whether a change is truly a transformation. This method also provides a systematic process to forecast, understand, and

Strategy map perspective	Measureable outcomes	Strategic objectives	Organizational transformation
Financial	• Improve returns on project investments • Reduce technology overhead costs	• Combine IT expenses with relocation and capitalize entire expense • Integrate new telephone system with computer network expenses • Leverage engineering and communications expenses with technology • Retire old equipment from financial statements	Combination of expenses requires formation of new communities of practice, which includes finance, engineering, and IT
Users	• More satisfied users • Increase user IT support	• Increase access to central applications • Integrate IT within other departments to improve dynamic customer support requirements • Provide new products to replace old e-mail system and make standard applications available to all users • Establish help desk personnel	Process of supporting users requires IT staff to embrace reflective practices. User relationship formed through new communities of practice and cultural assimilation with user community New culture at Ravell established
Process	• Provide accurate and timely information • Improved systems	• Improve decision support for improved reporting and strategic marketing • Upgrade new internal systems, including customer relationship management (CRM), general ledger, and rights and royalties	Startegic integration occurs through increased discourse and language among communities of practice engaged in making relocation successful. New knowledge created and needs knowledge management
Learning and growth	• New technology products • New ways of staff interaction structure • Establish new organization	• Investigate new voice-messaging technology to improve integration of e-mail and telephone systems • Physically relocate IT staff across departments • Modify IT reporting structure with "dotted line" to business units	IT becomes more critically reflective, understands value of their participation with learning organization. IT staff seeks to know less and understands view of the "other"

Figure 6.8 Ravell phase I balanced scorecard.

present what technology initiatives will ultimately change in the strategic integration and cultural assimilation components of ROD.

There are two other important factors embedded in this modified balanced scorecard technique. First, scorecards can be designed at varying levels of detail. Thus, two more balanced scorecards could be developed that reflect the organizational transformations that occurred in Ravell phases II and III, or the three phases could be summarized as one large balanced scorecard or some combination of summary and detail together. Second, the scorecard can be modified to reflect unexpected changes during implementation of a technology. These changes could be related to a shifting mission statement or to external changes in the market that require a change in business strategy. Most important, though, are the expected outcomes and transformations that occur during the course of a project. Essentially, it is difficult to predict how organizations will actually react to changes during an IT project and transform.

The balanced scorecard provides a checklist and tracking system that is structured and sustainable—but not perfect. Indeed, many of the outcomes from the three phases of Ravell were unexpected or certainly not exactly what I expected. The salient issue here is that it allows an organization to understand when such unexpected changes have occurred. When this does happen, organizations need to have an infrastructure and a structured system to examine what a change in their mission, strategy, or expectations means to all of the components of the project. This can be described as a "rippling effect," in which one change can instigate others, affecting many other parts of the whole. Thus, the balanced scorecard, particularly using a strategy map, allows practitioners to reconcile how changes will affect the entire plan.

Another important component of the balanced scorecard, and the reason why I use it as the measurement model for outcomes, is its applicability to organizational learning. In particular, the learning and growth perspective shows how the balanced scorecard ensures that learning and strategy are linked in organizational development efforts.

Implementing balanced scorecards is another critical part of the project—who does the work, what the roles are, and who has the responsibility for operating the scorecards? While many companies use consultants to guide them, it is important to recognize that balanced scorecards reflect the unique features and functions of the company. As such, the rank and file need to be involved with the design and support of balanced scorecards.

Every business unit that has a scorecard needs to have someone assigned to it, someone accountable for it. A special task force may often be required to launch the training for staff and to agree on how the scorecard should be designed and supported. It is advisable that the scorecard be implemented using some application software and made available on an Internet network. This provides a number of benefits:

> It reduces paper or local files that might get lost or not be secured, allows for easy "roll-up" of multiple scorecards, to a summary level, and access via the Internet (using an external secured hookup) allows the scorecard to be maintained from multiple locations. This is particularly attractive for staff members and management individuals who travel.

According to Olve et al. (2003), there are four primary responsibilities that can support balanced scorecards:

1. *Business stakeholders*: These are typically senior managers who are responsible for the group that is using the scorecard. These individuals are advocates of using scorecards and require compliance if deemed necessary. Stakeholders use scorecards to help them manage the life cycle of a technology implementation.
2. *Scorecard designers*: These individuals are responsible for the "look and feel" of the scorecard as well as its content. To some extent, the designers set standards for appearance, text, and terminology. In certain situations, the scorecard designers have dual roles as project managers. Their use of scorecards helps them understand how the technology will operate.
3. *Information providers*: These people collect, measure, and report on the data in the balanced scorecard. This function can be implemented with personnel on the business unit level or from a central services department. Reporting information often requires support from IT staff, so it makes sense to have someone from IT handle this responsibility. Information providers use the scorecard to perform the measurement of project performance and the handling of data.
4. *Learning pilots*: These individuals link the scorecard to organizational learning. This is particularly important when measuring organizational transformation and individual development.

The size and complexity of an organization will ultimately determine the exact configuration of roles and responsibilities that are needed to implement balanced scorecards. Perhaps the most applicable variables are:

Competence: Having individuals who are knowledgeable about the business and its processes, as well as knowledgeable about IT.

Availability: Individuals must be made available and appropriately accommodated in the budget. Balanced scorecards that do not have sufficient staffing will fail.

Executive management support: As with most technology projects, there needs to be a project advocate at the executive level.

Enthusiasm: Implementation of balanced scorecards requires a certain energy and excitement level from the staff and their management. This is one of those intangible, yet invaluable, variables.

Balanced Scorecards and Discourse

In Chapter 4, I discussed the importance of language and discourse in organizational learning. Balanced scorecards require ongoing dialogues that need to occur at various levels and between different communities of practice. Therefore, it is important to integrate language and discourse and communities of practice theory with balanced scorecard strategy. The target areas are as follows:

• Developing of strategy maps
• Validating links across balanced scorecard perspectives
• Setting milestones
• Analyzing results
• Evaluating organizational transformation

Figure 6.9 indicates a community of practice relationship that exists at a company. Each of these three levels was connected by a concept I called "common threads of communication." This model can be extended to include the balanced scorecard.

The first level of discourse occurs at the executive community of practice. The executive management team needs to agree on the

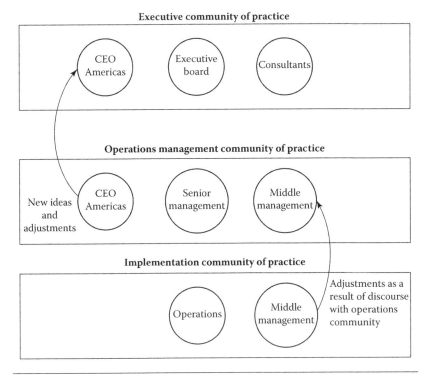

Figure 6.9   Community of practice "threads."

specific business strategy that will be used as the basis of the mission statement for the balanced scorecard. This requires conversations and meetings that engage the CEO, executive board members (when deemed applicable), and executive managers, like the chief operating officer (COO), chief financial officer (CFO), chief information officer (CIO), and so on. Each of these individuals needs to represent his or her specific area of responsibility and influence from an executive perspective. The important concept is that the balanced scorecard mission and strategy should be a shared vision and responsibility for the executive management team as a whole. To accomplish this task, the executive team needs to be instructed on how the balanced scorecard operates and on its potential for accomplishing organizational transformation that leads to strategic performance. Ultimately, the discourse must lead to a discussion of the four balanced scorecard perspectives: financial, customer, process, and learning and growth.

From a middle management level, the balanced scorecard allows for a measurable model to be used as the basis of discourse with executives. For example, the strategy map can be the vehicle for conducting meaningful conversations on how to transform execu-tive-level thinking and meaning into a more operationally focused strategy. Furthermore, the scorecard outlines the intended outcomes for strategy and organizational learning and transformation.

The concept of using the balanced scorecard as a method with which to balance thinking and meaning across communities of prac-tice extends to the operational level as well. Indeed, the challenge of making the transition from thinking and meaning at the executive level of operations is complicated, especially since these communi-ties rarely speak the same language. The measurable outcomes section of the scorecard provides the concrete layer of outcomes that opera-tions staff tend to embrace. At the same time, this section provides corresponding strategic impact and organizational changes needed to satisfy business strategies set by management.

An alternative method of fostering the need forms of discourse is to create multiple-tiered balanced scorecards designed to fit the language of each community of practice, as shown in Figure 6.10. The diagram in Figure 6.10 shows that each community can maintain its own lan-guage and methods while establishing "common threads" to foster a transition of thinking and meaning between it and other communi-ties. The common threads from this perspective look at communica-tion at the organizational/group level, as opposed to the individual level. This relates to my discussion in Chapter 4, which identified individual methods of improving personal learning, and development within the organization. This suggests that each balanced scorecard must embrace language that is common to any two communities to establish a working and learning relationship—in fact, this common language is the relationship.

Knowledge Creation, Culture, and Strategy

Balanced scorecards have been used as a measurement of knowledge creation. Knowledge creation, especially in technology, has signifi-cant meaning, specifically in the relationship between data and infor-mation. Understanding the sequence between these two is interesting. We know that organizations, through their utilization of software applications, inevitably store data in file systems called databases.

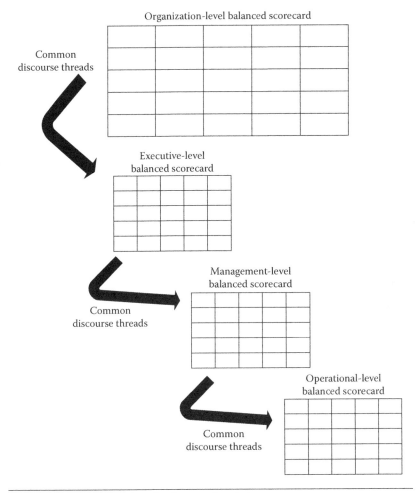

Figure 6.10 Community of practice "common threads."

The information stored in these databases can be accessed by many different software applications across the organization. Accessing multiple databases and integrating them across business units creates further valuable information. Indeed, the definition of information is "organized data." These organized data are usually stored in data infrastructures called data warehouses or data marts, where the information can be queried and reported on to assist managers in their decision-making processes. We see, in the Ravell balanced scorecard, that decision-support systems were actually one of the strategic objectives for the process perspective.

Unfortunately, information does not ensure new knowledge creation. New knowledge can only be created by individuals who evolve in their roles and responsibilities. Individuals, by participating in groups and communities of practice, can foster the creation of new organizational knowledge. However, to change or evolve one's behavior, there must be individual or organizational transformation. This means that knowledge is linked to organizational transformation. The process to institutionalize organizational transformation is dependent on management interventions at various levels. Management needs to concentrate on knowledge management and change management and to act as a catalyst and advocate for the successful implementation of organizational learning techniques. These techniques are necessary to address the unique needs of ROD.

Ultimately, the process must be linked to business strategy. ROD changes the culture of an organization, through the process of cultural assimilation. Thus, there is an ongoing need to reestablish alignment between culture and strategy, with culture altered to fit new strategy, or strategy first, then culture (Pietersen, 2002). We see this as a recurring theme, particularly from the case studies, that business strategy must drive organizational behavior, even when technology acts as a dynamic variable. Pietersen identifies what he called six myths of corporate culture:

1. Corporate culture is vague and mysterious.
2. Corporate culture and strategy are separate and distinct things.
3. The first step in reducing our company should be defining our values.
4. Culture cannot be measured or rewarded.
5. Our leaders must communicate what our culture is.
6. Our culture is the one constant that never changes.

Resulting from these myths, Pietersen (2002) establishes four basic rules of success for creating a starting point for the balance between culture and strategy:

1. Company values should directly support strategic priorities.
2. They should be described as behaviors.
3. They should be simple and specific.

4. They should be arrived at through a process of enrollment (motivation).

Once business synergy is created, sustaining the relationship becomes an ongoing challenge. According to Pietersen (2002), this must be accomplished by continual alignment, measurement, setting examples, and a reward system for desired behaviors. To lead change, organizations must create compelling statements of the case for change, communicate constantly and honestly with their employees, maximize participation, remove ongoing resistance in the ranks, and generate some wins. The balanced scorecard system provides the mechanism to address the culture–strategy relationship while maintaining an important link to organizational learning and ROD. These linkages are critical because of the behavior of technology. Sustaining the relationship between culture and strategy is simply more critical with technology as the variable of change.

Ultimately, the importance of the balanced scorecard is that it forces an understanding that everything in an organization is connected to some form of business strategy. Strategy calls for change, which requires organizational transformation.

Mission: To accelerate investment in technology during the relocation of the company for reasons of economies of scale and competitive advantage.

7

Virtual Teams and Outsourcing

Introduction

Much has been written and published about virtual teams. Most define virtual teams as those that are geographically dispersed, although others state that virtual teams are those that primarily interact electronically. Technology has been the main driver of the growth of virtual teams. In fact, technology organizations, due mostly to the advent of competitive outsourcing abroad, have pushed information technology (IT) teams to learn how to manage across geographical locations, in such countries as India, China, Brazil, Ireland, and many others. These countries are not only physically remote but also present barriers of culture and language. These barriers often impede communications about project status, and affect the likelihood of delivering a project on time, and within forecasted budgets.

Despite these major challenges, outsourcing remains attractive due to the associated cost savings and talent supply. These two advantages are closely associated. Consider the migration of IT talent that began with the growth of India in providing cheap and educated talent. The promise of cost savings caused many IT development departments to begin using more India-based firms. The ensuing decline in IT jobs in the United States resulted in fewer students entering IT curriculums at U.S. universities for fear that they would not be able to find work. Thus, began a cycle of lost jobs in the United States and further demand for talent abroad. Now, technology organizations are faced with the fact that they *must* learn to manage virtually because the talent they need is far away.

From an IT perspective, successful outsourcing depends on effective use of virtual teams. However, the converse is not true; that is, virtual teams do not necessarily imply outsourcing. Virtual teams can be made up of workers anywhere, even those in the United States who are working from a distance rather than reporting to an office for work. A growing number of employees in the United States want more personal flexibility; in response, many companies are allowing employees to work from home more often—and have found the experience most productive. This type of virtual team management generally follows a hybrid model, with employees working at home most of the time but reporting to the office for critical meetings; an arrangement that dramatically helps with communication and allows management to have quality checkpoints.

This chapter addresses virtual teams working both within the United States and on an outsource basis and provides readers with an understanding of when and how to consider outsource partners. Chapter topics include management considerations, dealing with multiple locations, contract administration, and in-house alternatives. Most important, this chapter examines organizational learning as a critical component of success in using virtual teams. Although the advent of virtual teams creates another level of complexity for designing and maintaining learning organizations, organizational learning approaches represent a formidable solution to the growing dilemma of how teams work, especially those that are 100% virtual.

Most failures in virtual management are caused by poor communication. From an organizational learning perspective, we would define this as differences in meaning making—stemming mostly from cultural differences in the meaning of words and differing behavioral norms. There is also no question that time zone differences play a role in certain malfunctions of teams, but the core issues remain communication related.

As stated, concerning the Ravell case study, cultural transformation is slow to occur and often happens in small intervals. In many virtual team settings, team members may never do more than communicate via e-mail. As an example, I had a client who was outsourcing production in China. One day, they received an e-mail stating, "We cannot do business with you." Of course, the management team was confused and worried, seeking to understand why the business arrangement was ending without any formal discussions of the problem. A translator in China was hired to help clarify the dilemma. As it turned out, the statement was meant to suggest that the company needed to provide more business—more work, that is. The way the Chinese communicated that need was different from the Western interpretation. This is just a small example of what can happen without a well-thought-out organizational learning scheme. That is, individuals need to develop more reflective abilities to comprehend the meaning of words before they take action, especially in virtual environments across multiple cultures. The development of such abilities—the continual need for organizations to respond effectively to dynamic changes, brought about by technology, in this case, e-mail—is consistent with my theory of responsive organizational dynamism (ROD). The e-mail established a new dynamic of communication. Think how often specifications and product requirements are changing and need virtual teams to somehow come together and agree on how to get the work done—or think they agree.

Prior research and case studies provide tools and procedures as ways to improve productivity and quality of virtual team operations. While such processes and methodologies are helpful, they will not necessarily ensure the successful outcomes that IT operations seek unless they also change. Specifically, new processes alone are not sufficient or a substitute for learning how to better communicate and make meaning in a virtual context. Individuals must learn how to develop new behaviors when working virtually. We must also remember that virtual team operations are not limited to IT staffs. Business users often need to be involved as they would in any project, particularly when users are needed to validate requirements and test the product.

Status of Virtual Teams

The consensus tells us that virtual teams render results. According to Bazarova and Walther (2009), "Virtual groups whose members communicate primarily or entirely via email, computer conferencing, chat, or voice—have become a common feature of twenty-first century organizations" (p. 252). Lipnack and Stamps (2000) state that virtual teams will become the accepted way to work and will likely reshape the work world. While this prediction seems accurate, there has also been evidence of negative attribution or judgment about problems that arise in virtual team performance. Thus, it is important to understand how virtual teams need to be managed and how realistic expectations of such teams might be formed. So, while organizations understand the need for virtual teams, they are not necessarily happy with project results. Most of the disappointment relates to a lack of individual development that helps change the mindset of how people need to communicate, coupled with updated processes.

Management Considerations

Attribution theory "describes how people typically generate explanations for outcomes and actions—their own and others" (Bazarova & Walther, 2009, p. 153). This theory explains certain behavior patterns that have manifested during dysfunctional problems occurring in managing virtual teams. Virtual teams are especially vulnerable to such problems because their limited interactions can lead to members not having accurate information about one another. Members of virtual teams can easily develop perceptions of each other's motives that are inaccurate or distorted by differing cultural norms. Research also shows us that virtual team members typically attribute failure to the external factors and successes to internal factors. Problems are blamed on the virtual or outside members for not being available or accountable to the physical community. The successes then tend to reinforce that virtual teams are problematic because of their very nature. This then establishes the dilemma of the use of virtual teams and organizations—its use will continue to increase and dominate workplace structures and yet will present challenges to organizations that do not want to change. The lack of support to change will be substantiated during failures in expected outcomes. Some of the failures, however, can and should be attributable to distance. As Olson and Olson (2000) state: "Distance will persist as an important element of human experience" (p. 172). So, despite the advent of technology, it is important not to ignore the social needs that teams need to have to be effective.

Dealing with Multiple Locations

Perhaps the greatest difficulty in implementing virtual teams is the reality that they span multiple locations. More often, these locations can be in different time zones and within multiple cultures. To properly understand the complexity of interactions, it makes sense to revisit the organizational learning tools discussed in prior chapters. Perhaps another way of viewing virtual teams and their effects on organization learning is to perceive it as another dimension—a dimension that is similar to multiple layers in a spreadsheet. This notion means that virtual teams do not upset the prior relations between technology as a variable from a two-dimensional perspective, rather in the depth of how it affects this relationship in a third dimension. Figure 7.1 reflects how this dimension should be perceived.

In other words, the study of virtual teams should be viewed as a subset of the study of organizations. When we talk about workplace activities, we need to address issues at the component level. In this example, the components are the physical organization and the

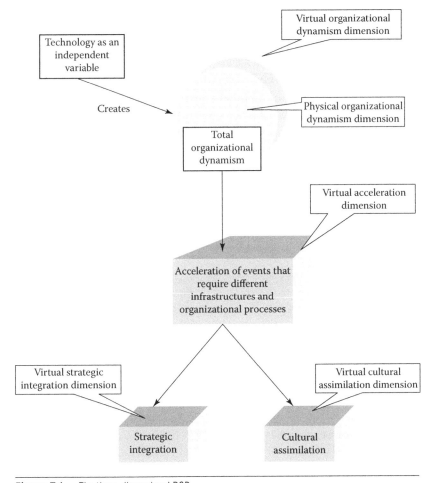

Figure 7.1   The three-dimensional ROD.

virtual organization. The two together make up the superset or the entire organization. To be fruitful, any discussion of virtual organizations must be grounded in the context of the entire organization and address the complete topic of workplace learning and transformation. In Chapter 4, I discussed organizational learning in communities of practice (COP). In this section, I expand that discussion to include virtual organizational structures.

The growing use of virtual teams may facilitate the complete integration of IT and non-IT workers. The ability to connect from various locations using technology itself has the potential to expand COP. But, as discussed in Chapter 4, it also presents new challenges, most of which relate to the transient nature of members, who tend to participate on more of a subject or transactional basis, rather than being permanent members of a group. Table 7.1 reflects some of the key differences between physical and virtual teams.

There has been much discussion about whether every employee is suited to perform effectively in a virtual community. The consensus is that effective virtual team members need to be self-motivated, able to work independently, and able to communicate clearly and in a positive way. However, given that many workers lack some or all of these skills, it seems impractical to declare that workers who do not meet these criteria should be denied the opportunity to work in virtual

Table 7.1   Operating Differences between Traditional and Virtual Teams

TRADITIONAL OR PHYSICAL TEAMS	VIRTUAL TEAMS
Teams tend to have fixed participation and members.	Membership shifts based on topics and needs.
Members tend to be from the same organization.	Team members can include people from outside the organization (clients and collaborators).
Team members are 100% dedicated.	Members are assigned to multiple teams.
Team members are collocated geographically and by organization.	Team members are distributed geographically and by organization.
Teams tend to have a fixed term of membership; that is, start and stop dates.	Teams are reconfigured dynamically and may never terminate.
Teams tend to have one overall manager.	Teams have multiple reporting relationships with different parts of the organization at different times.
Teamwork is physical and practiced in face-to-face interactions.	Teamwork is basically social.
Engagement is often during group events and can often be hierarchical in nature.	Individual engagement is inseparable from empowerment.

teams. A more productive approach might be to encourage workers to recognize that they must adapt to changing work environments at the risk of becoming marginal in their organizations.

To better understand this issue, I extended the COP matrix, presented in Chapter 4, to include virtual team considerations in Table 7.2.

Item 7 in Table 7.2 links the study of knowledge management with COP. Managing knowledge in virtual communities within an organization has become associated directly with the ability of a firm to sustain competitive advantage. Indeed, Peddibhotla and Subramani (2008) state that "virtual communities are not only recognized as important contributors to both the development of social networks among individuals but also towards individual performance and firm performance" (p. 229). However, technology-enabled facilities and support, while providing a repository for better documentation, also create challenges in maintaining such knowledge. The process of how information might become explicit has also dramatically changed with the advent of virtual team communications. For example, much technology-related documentation evolves from bottom-up sources, rather than the traditional top-down process. In effect, virtual communities share knowledge more on a peer-to-peer basis or through mutual consensus of the members. As a result, virtual communities have historically failed to meet expectations, particularly those of management, because managers tend to be uninvolved in communication. While physical teams can meet with management more often before making decisions, virtual teams have no such contact available. To better understand the complexities of knowledge management and virtual teams, Sabherwal and Becerra-Fernandez (2005) expand on Nonaka's (1994) work on knowledge management, which outlined four modes of knowledge creation: externalization, internalization, combination, and socialization. Each of these modes is defined and discussed next.

Externalization

Externalization is the process of converting or translating tacit knowledge (undocumented knowledge) into explicit forms. The problem with this concept is whether individuals really understand what they know

Table 7.2 Communities of Practice: Virtual Team Extensions

STEP	COMMUNITIES-OF-PRACTICE STEP	TECHNOLOGY EXTENSION	VIRTUAL EXTENSION
1	Understanding strategic knowledge needs: What knowledge is critical to success.	Understanding how technology affects strategic knowledge and what specific technological knowledge is critical to success.	Understanding how to integrate multiple visions of strategic knowledge and where it can be found across the organization.
2	Engaging practice domains: Where people form communities of practice to engage in and identify with.	Technology identifies groups based on business-related benefits, requiring domains to work together toward measurable results.	Virtual domains are more dynamic and can be formed for specific purposes and then reconfigured based on practice needs of subjects discussed.
3	Developing communities: How to help key communities reach their full potential.	Technologies have life cycles that require communities to continue; treats the life cycle as a supporter for attaining maturation and full potential.	Communities can be reallocated to participate in multiple objectives. Domains of discussion have no limits to reach organizational needs.
4	Working the boundaries: How to link communities to form broader learning systems	Technology life cycles require new boundaries to be formed. This will link other communities that were previously outside of discussions and thus expand input into technology innovations.	Virtual abilities allow for customer interfaces, vendors, and other interested parties to join the community.
5	Fostering a sense of belonging: How to engage people's identities and sense of belonging.	The process of integrating communities: IT and other organizational units will create new evolving cultures that foster belonging as well as new social identities.	Communities establish belonging in a virtual way. Identities are established more on content of discussion than on physical attributes of members.
6	Running the business: How to integrate communities of practice into running the business of the organization.	Cultural assimilation provides new organizational structures that are necessary to operate communities of practice and to support new technological innovations.	The organization functions more as a virtual community or team, being more agile to demands of the business, and interactions may not always include all members.

(*Continued*)

Table 7.2 (Continued)   Communities of Practice: Virtual Team Extensions

STEP	COMMUNITIES-OF-PRACTICE STEP	TECHNOLOGY EXTENSION	VIRTUAL EXTENSION
7	Applying, assessing, reflecting, renewing: How to deploy knowledge strategy through waves of organizational transformation.	The active process of dealing with multiple new technologies that accelerates the deployment of knowledge strategy. Emerging technologies increase the need for organizational transformation.	Virtual systems allow for more knowledge strategy because of the ability to deploy information and procedures. Tacit knowledge is easier to transform to explicit forms.

and how it might affect organizational knowledge. Virtual communities have further challenges in that the repository of tacit information can be found in myriad storage facilities, namely, audit trails of e-mail communications. While Sabherwal and Becerra-Fernandez (2005) suggest that technology may indeed assist in providing the infrastructure to access such information, the reality is that the challenge is not one of process but rather of thinking and doing. That is, it is more a process of unlearning existing processes of thinking and doing, into new modes of using knowledge that is abundantly available.

Internalization

Internalization is a reversal of externalization: It is the process of transferring explicit knowledge into tacit knowledge—or individualized learning. The individual thus makes the explicit process into his or her own stabilized thinking system so that it becomes intuitive in operation. The value of virtual team interactions is that they can provide more authentic evidence of why explicit knowledge is valuable to the individual. Virtual systems simply can provide more people who find such knowledge useful, and such individuals, coming from a more peer relationship, can understand why their procedures can be internalized and become part of the self.

Combination

Combination allows individuals to integrate their physical processes with virtual requirements. The association, particularly in a global environment, allows virtual team members to integrate new explicit forms into their own, not by replacing their beliefs, but rather, by establishing new hybrid knowledge systems. This is particularly advantageous across multiple cultures and business systems in countries that hold different and, possibly complementary, knowledge about how things can get done. Nonaka's (1994) concept of combination requires participants in the community to be at later stages of multiplicity—suggesting that this form can only be successful among certain levels or positions of learners.

Socialization

As Nonaka (1994) notes, individuals learn by observation, imitation, and practice. The very expansion of conversations via technology can provide a social network in which individuals can learn simply through discourse. Discourse, as I discussed in Chapter 4, is the basis of successful implementations of COP. The challenge in virtual social networks is the difficulty participants have in assessing the authenticity of the information provided by those in the community.

The four modes of knowledge management formulated by Nonaka (1994) need to be expanded to embrace the complexities of virtual team COPs. Most of the adjustments are predicated on the team's ability to deal with the three fundamental factors of ROD that I introduced in this book; that is, acceleration, dynamic, and unpredictability. The application of these three factors of ROD to Nonaka's four modes is discussed next.

Externalization Dynamism

The externalization mode must be dynamic and ongoing with little ability to forecast the longevity of any tacit-to-explicit formulation. In other words, tacit-to-explicit change may occur daily but may only operate effectively for a shorter period due to additional changes brought on by technology dynamism. This means that members in a community must continually challenge themselves to revisit previous tacit processes and acknowledge the need to reformulate their tacit systems. Thus, transformation from tacit knowledge to explicit knowledge can be a daily challenge for COP virtual organizations.

Internalization Dynamism

Careful reflection on this process of internalizing explicit forms must be done. Given the differences in cultures and acceleration of business change, individualized learning creating new tacit abilities may not operate the same in different firm settings. It may be necessary to adopt multiple processes depending on the environment in which tacit operations are being performed. As stated, what might work in China may not work in Brazil, for example. Tacit behavior is culture oriented, so multiple and simultaneous versions must be respected and practiced. Further expansion of internalization is a virtual team's understanding of how such tacit behaviors change over time due to the acceleration of new business challenges.

Combination Dynamism

I believe the combination dynamism mode is the most important component of virtual team formation. Any combination or hybrid model requires a mature self—as specified in my maturity arcs discussed in Chapter 4. This means that individuals in virtual teams may need to be operating at a later stage of maturity to deal with the complexities of changing dispositions. Members of COPs must be observed, and a determination of readiness must be made for such new structures to develop in a virtual world. Thus, COP members need training; the lack of such training might explain why so many virtual teams have had disappointing results. Readiness for virtual team participation depends on a certain level of relativistic thinking. To be successful, virtual team members must be able to see themselves outside their own world and have the ability to understand the importance of what "others" need. This position suggests that individuals need to be tested to the extent that they are ready for such challenges. Organizational learning techniques remain a valid method for developing workers who can cope with the dynamic changes that occur in virtual team organizations.

Socialization Dynamism

Socialization challenges the virtual team members' abilities to understand the meaning of words and requires critical reflection of its constituents. ROD requires that virtual teams be agile and, especially, that they be responsive to the emotions of others in the community. This may require individuals to understand another member's maturity. Thus, virtual team members need to be able to understand why another member is behaving as he or she is or reacting in a dualistic manner. Assessment in a virtual collaboration becomes a necessity, especially given the unpredictability of technology-based projects.

In Table 5.1, I showed how tacit knowledge is mapped to ROD. Table 7.3 further extends this mapping to include virtual teams.

The requirements support research findings that knowledge management in a virtual context has significant factors that must be addressed to improve its success. These factors include management commitment, resource availability, modification of work practices, marketing of the initiative, training, and facilitation of cultural differences (Peddibhotla & Subramani, 2008).

The following are some action steps that organizations need to take to address these factors:

1. The executive team needs to advocate the commitment and support for virtual teams. The chief information officer (CIO) and his or her counterparts need to provide teams with the "sponsorship" that the organization will need to endure setbacks until the virtual organization becomes fully integrated into the learning organization. This commitment can be accomplished via multiple actions, including, but not limited to, a kickoff meeting with staff, status reports to virtual teams on successes and setbacks, e-mails and memos on new virtual formations, and a general update on the effort, perhaps on a quarterly basis. This approach allows the organization to understand the evolution of the effort and know that virtual teams are an important direction for the firm.

2. There should be training and practice sessions with collocated groups that allow teams to voice their concerns and receive direction on how best to proceed. Practice sessions should focus on team member responsibilities and advocating their ownership of responsibility. These sessions should cover lessons learned from actual experiences, so that groups can learn

Table 7.3   Tacit Knowledge and Virtual Teams

TACIT KNOWLEDGE	STRATEGIC INTEGRATION	STRATEGIC INTEGRATION VIRTUAL	CULTURAL ASSIMILATION	CULTURAL ASSIMILATION VIRTUAL
Cultural and social history			How the IT department and other departments translate emerging technologies into their existing processes and organization.	How can virtual and nonvirtual departments translate emerging technologies into their projects across multiple locations and cultures?
Problem-solving modes	Individual reflective practices that assist in determining how specific technologies can be useful and how they can be applied; utilization of tacit knowledge to evaluate probabilities for success.	Individual reflective practices and intercultural communications needed to determine how tacit knowledge should be applied to specific group and project needs.	Technology opportunities may require organizational and structural changes to transfer tacit knowledge to explicit knowledge.	Technological opportunities may require configuration of virtual communities of practice and explicit knowledge.
Orientation to risks and uncertainties	Technology offers many risks and uncertainties. All new technologies may not be valid for the organization. Tacit knowledge is a valuable component to fully understand realities, risks, and uncertainties.	Technology risks and uncertainties need to be assessed by multiple virtual and physical teams to determine how technologies will operate across multiple locations and cultures.		

(Continued)

Table 7.3 (Continued)   Tacit Knowledge and Virtual Teams

TACIT KNOWLEDGE	STRATEGIC INTEGRATION	STRATEGIC INTEGRATION VIRTUAL	CULTURAL ASSIMILATION	CULTURAL ASSIMILATION VIRTUAL
Worldviews			Technology has global effects and changes market boundaries, that cross business cultures; it requires tacit knowledge to understand existing dispositions on how others work together. Reviews how technology affects the dynamics of operations.	Market boundaries are more dynamic across virtual teams that operate to solve cross-cultural and business problems. Worldviews are more the norm than the exception.
Organizing principles			How will new technologies actually be integrated? What are the organizational challenges to "rolling out" products, and to implementation timelines? What positions are needed, and who in the organization might be best qualified to fill new responsibilities? Identify limitations of the organization; that is, tacit knowledge versus explicit knowledge realities.	What are the dynamic needs of the virtual team, to handle new technologies on projects? What are the new roles and responsibilities of virtual team members? Determine what tacit and explicit knowledge will be used to make decisions.
Horizons of expectation	Individual limitations in the tacit domain that may hinder or support whether a technology can be strategically integrated into the organization.	Individuals within the virtual community need to understand the limitations on strategic uses of technology. This may vary across cultures.		

from others. Training should set the goals and establish the criteria for how virtual teams interact in the firm. This should include the application software and repositories that are in place and the procedures for keeping information and knowledge current.

3. External reminders should be practiced so that virtual teams do not become lax and develop bad habits since no one is monitoring or measuring success. Providing documented processes, perhaps a balanced scorecard or International Organization for Standardization (ISO) 9000-type procedures and measurements, is a good practice for monitoring compliance.

Dealing with Multiple Locations and Outsourcing

Virtual organizations are often a given in outsourcing environments, especially those that are offshore. Offshore outsourcing also means that communications originate in multiple locations. The first step in dealing with multiple locations is finding ways to deal with different time zones. Project management can become more complicated when team meetings occur at obscure times for certain members of the community. Dealing with unanticipated problems can be more challenging when assembling the entire team may not be feasible because of time differences. The second challenge in running organizations in multiple locations is culture. Differing cultural norms can especially cause problems during off-hour virtual sessions. For example, European work culture does not often support having meetings outside work hours. In some countries, work hours may be regulated by the government or powerful unions.

A further complication in outsourcing is that the virtual team members may be employed by different companies. For instance, part of the community may include a vendor who has assigned staff resources to the effort. Thus, these outsourced team members belong to the community of the project yet also work for another organization. The relationship between an outside consultant and the internal team is not straightforward and varies among projects. For example, some outsourced technical resources may be permanently assigned to the project, so while they actually work for another firm, they behave and take daily direction as if they were an employee of the focal business. Yet, in other relationships, outsourced resources work closely under the auspices of the outsourced "project manager," who acts as a buffer between the firm and the vendor. Such COP formations vary. Still other outsourcing arrangements involve team members the firm does not actually know unless outsourced staff is called in to solve a problem. This situation exists when organizations outsource complete systems, so that the expectation is based more on the results than on the interaction. Notwithstanding the arrangement or level of integration, a COP must exist, and its behavior in all three of these examples varies in participation, but all are driven in a virtual relationship more by dynamic business events than by preplanned activities.

If we look closely at COP approaches to operations, it is necessary to create an extension of dynamism in a virtual team community. The extension reflects the reliance on dynamic transactions, which creates temporary team formations based on demographic similarity needs. This means that virtual teams will often be formed based on specific interests of people within the same departments. Table 7.4 shows the expansion of dynamism in a virtual setting of COPs.

Thus, the advent of modern-day IT outsourcing has complicated the way COPs function. IT outsourcing has simultaneously brought attention to the importance of COP and knowledge management in general. It also further supports the reality of technology dynamism as more of a norm in human communication in the twenty-first century.

Revisiting Social Discourse

In Chapter 4, I covered the importance of social discourse and the use of language as a distinct component of how technology changes COP. That section introduced three components that linked talk and action, according to the schema of Grant et al. (1998): Identity, skills and emotion. Figure 7.2 shows this relationship again. The expansion of virtual team communications further emphasizes the importance of discourse and the need to rethink how these three components relate to each other in a virtual context.

Table 7.4   COP Virtual Dynamism

COP PHYSICAL SOCIAL SETTINGS	COP VIRTUAL DYNAMISMS
There is shared pursuit of interest accomplished through group meetings.	Interest in discussion is based more on dynamic transactions and remote needs to satisfy specific personal needs.
Creation of the "community" is typically established within the same, or similar, departments.	The notion of permanency is deemphasized. Specific objectives based on the needs of the group will establish the community.
Demographic similarity is a strong contributor to selection of community members.	Demographic similarity has little to do with community selection. Selection is based more on subject-matter expertise.
Situated learning is often accomplished by assisting members to help develop others. Learning occurs within a framework of social participation.	Situated learning to help others has less focus. It may not be seen as the purpose or responsibility of virtual team members. Social participation has more concrete perspective.
Community needs to assess technology dynamism using ROD in more physical environments requiring a formal infrastructure.	Community is less identifiable from a physical perspective. ROD must be accomplished by members who have special interests at the subject level as opposed to the group level.
COP works well with cultural assimilation of formal work groups where participants are clearly identified.	Cultural assimilation in virtual settings is more transaction-based. Assimilation can be a limited reality during the time of the transaction to ensure success of outcomes.
COP can be used for realignment of work departments based on similar needs.	COP in a virtual environment creates temporary realignments, based on similar needs during the process.
COP supports continual learning and dealing with unplanned action.	COPs are continually reconfigured, and do not have permanency of group size or interest.

Identity

I spoke about the "cultures of practice" due to expansion of contacts from technology capacities. This certainly holds true with virtual teams. However, identities can be transactional—in ways such that an individual may be a member of multiple COP environments and have different identities in each. This fact emphasizes the multitasking aspect of the linear development modules discussed throughout this book. Ultimately, social discourse will dynamically change based on the COP to which an individual belongs, and that individual needs to be able to "inventory" these multiple roles and responsibilities. Such roles and responsibilities themselves will transform, due to the dynamic nature of technology-driven projects. Individuals will thus have multiple identities and must be able to manage those identities across different COPs and in different contexts within those COPs.

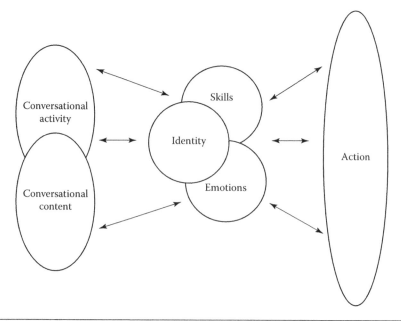

Figure 7.2   Grant's schema of the relationship between talk and action.

This requires individual maturities that must be able to cope with the "other" and understand the relativistic nature of multiple cultures and the way discourse transforms into action.

Skills

I mentioned the importance of persuasion as a skill to transform talk into action. Having the ability to persuade others across virtual teams is critical. Often, skills are misrepresented as technical abilities that give people a right of passage. Across multiple cultures, individuals in teams must be able to recognize norms and understand how to communicate with others to get tangible results on their projects. It is difficult to make such determinations about individuals that one has never met face to face. Furthermore, virtual meetings may not provide the necessary background required to properly understand a person's skill sets, both "hard" and "soft." The soft skills analysis is more important as the individual's technical credentials become assumed. We see such assumptions when individuals transition into management positions. Ascertaining technical knowledge at the staff level is easier—almost like an inventory analysis of technical requirements.

However, assessing an individual's soft skills is much more challenging. Virtual teams will need to create more complex and broadened inventories of their team's skill sets, as well as establish better criteria on how to measure soft skills. Soft skills will require individuals to have better "multicultural" abilities, so that team members can be better equipped to deal with multinational and cross-cultural issues.

Emotion

Like persuasion, emotion involves an individual's ability to motivate others and to create positive energy. Many of those who successfully use emotion are more likely to have done so in a physical context than a virtual one. Transferring positive emotion in a virtual world can be analogous to what organizations experienced in the e-commerce world, in which organizations needed to rebrand themselves across the Web in such a way that their image was reflected virtually to their customers. Marketing had to be accomplished without exposure to the buyer during purchase decisions. Virtual COPs are similar: Representation must be what the individual takes away, without seeing the results physically. This certainly offers a new dimension for managing teams. This means that the development requirements for virtual members must include advanced abstract thinking so that the individual can better forecast virtual team reactions to what *will* be said, as opposed to reacting while the conversation is being conducted or thinking about what to do after virtual meetings.

In Chapter 4, I presented Marshak's (1998) work on types of talk that lead to action: tool-talk, frame-talk, and mythopoetic-talk. Virtual teams require modification to the sequence of talk; that is, the use of talk is altered. Let us first look at Figure 7.3, representing Marshak's model. To be effective, virtual teams must follow this sequence from the outside inward. That is, the virtual team must focus on mythopoetic-talk in the center as opposed to an outer ring. This means that ideogenic issues must precede interpretation in a virtual world. Thus, tool-talk, which in the physical world lies at the center of types of tools, is now moved to the outside rectangle. In other words, instrumental actions lag those of ideology and interpretation. This is restructured in Figure 7.4.

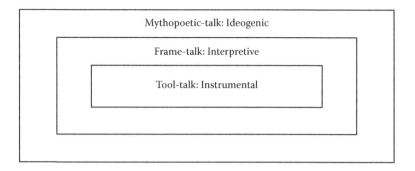

Figure 7.3   Marshak's types of talk.

Mythopoetic-talk is at the foundation of grounding ideas in a virtual COP. It would only make sense that a COP-driven talk requires ideogenic behavior before migrating to instrumental outcomes. Remember that ideogenic talk allows for concepts of intuition and ideas for concrete application especially relevant among multiple cultures and societies. So, we again see that virtual teams require changes in the sequence of how learning occurs. This change in sequence places more emphasis on the need for an individual to be more developmentally mature—with respect to thinking, handling differences, and thinking abstractly. This new "abstract individual" must be able to reflect before action and reflect in action to be functionally competent in virtual team participation.

Because ROD is relevant, it is important to determine how virtual teams affect the ROD maturity arc first presented in Figure 4.10 and redisplayed in Figure 7.5.

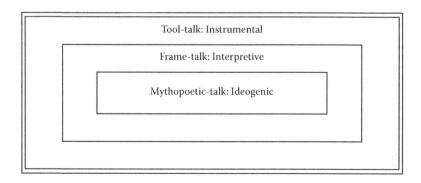

Figure 7.4   Virtual team depiction of Marshak's types of talk.

Stages of individual and organizational learning

Sector variable	Operational knowledge	Department/unit view as other	Integrated disposition	Stable operations	Organizational leadership
Strategic integration	Operations personnel understand that technology has an impact on strategic development, particularly on existing processes	Individual beliefs of strategic impact are incomplete; individual needs to incorporate other views within the department or business unit	Recognition that individual and department views must be integrated to be complete and strategically productive for the department/unit	Changes made to processes at the department/unit level formally incorporate emerging technologies	Department strategies are propagated and integrated at organization level
Cultural assimilation	View that technology can and will affect the way the organization operates and that it can affect roles and responsibilities	Changes brought forth by technology need to be assimilated into departments and are dependent on how others participate	Understands need for organizational changes; different cultural behavior new structures are seen as viable solutions	Organizational changes are completed and in operation; existence of new or modified employee positions	Department-level organizational changes and cultural evolution are integrated with organization-wide functions and cultures
Organizational learning constructs	Individual-based reflective practice	Small-group based reflective practices	Interactive with both individual and middle management using communities of practice	Interactive between middle management and executives using social discourse methods to promote transformation	Organizational learning at executive level using knowledge management
Management level	Operations	Operation and middle management	Middle management	Middle management and executive	Executive

Figure 7.5   Responsive organizational dynamism arc model.

Cultural assimilation	View that technology can and will affect the way the organization operates, and that it can affect roles and responsibilities	*Changes brought forth by technology need to be assimilated into departments and are dependent on how others participate. Assimilation of cultural norm may have very different roles and responsibilities in other cultures. Shifting assimilation needs may differ as different members join or leave the COP*	*Understands need for organizational changes across multiple organizations. Different cultural behaviors and new structures are seen as viable solutions that can be permanent or temporary because of the dynamic memberships in COP*	*Organizational changes are never completed and may be in temporary operation. Existence of new or modified COP member positions could be permanent or transitional based on project needs*	*Department-level organizational changes and cultural evolution may remain separate and case driven. Some assimilation may be integrated with organizaion-wide functions and cultures*
Organizational learning constructs	Individual-based reflective practice	*Virtual group-based reflective practices are necessary to understand how to operate in a COP environment with individuals who have different perspectives*	*Interactive with individual, middle management, and executive using virtual communities of practice*	*Interactive between middle management and executives using social discourse methods to promote more transactional behavioral transitions. Some transitions may lead to transformation*	*Organizational learning at executive level incorporates virtual team needs using more dynamic COP structures and broadened knowledge managment that is more situational*
Management level	Operations	*Operations, middle management, and executive*	*Middle management and executive*	*Middle management and executive*	*Executive*

Figure 7.6 Virtual team extension to the ROD arc. Changes are shown in italics.

Figure 7.6 represents the virtual team extension to the ROD arc. The changes to the cells are shown in italics. Note that there are no changes to operational knowledge because this stage focuses solely on self-knowledge learned from authoritative sources. However, as the individual matures, there is greater need to deal with uncertainty. This includes the uncertainty that conditions in a COP may be temporary, and thus knowledge may need to vary from meeting to meeting. Furthermore, while operational realities may be more transactional, it does not necessarily mean that adopted changes are not permanent. Most important is the reality that permanence in general may no longer be a characteristic of how the organization operates; this further emphasizes ROD as a way of life. As a result of this extreme complexity in operations, there is an accelerated requirement for executives to become involved earlier in the development process. Specifically, by stage two (department/unit view of the other), executives must be engaged in virtual team management considerations.

Ultimately, the virtual team ROD arc demonstrates that virtual teams are more complex and therefore need members who are more mature to ensure the success of outsourcing and other virtual constructs. It also explains why virtual teams have struggled, likely because their members are not ready for the complex participation necessary for adequate outcomes.

We must also remember that maturity growth is likely not parallel in its linear progression. This was previously shown in Figure 4.12.

This arc demonstrates the challenge managers face in gauging the readiness of their staff to cope with virtual team engagement. On the other hand, the model also provides an effective measurement schema that can be used to determine where members should be deployed and their required roles and responsibilities. Finally, the model allows management to prepare staff for the training and development they need as part of the organizational learning approach to dealing with ROD.

8

SYNERGISTIC UNION OF IT AND ORGANIZATIONAL LEARNING

Introduction

This chapter presents case studies that demonstrate how information technology (IT) and organizational learning occur in the real corporate world. It examines the actual processes of how technological and organizational learning can be implemented in an organization and what management perspectives can support its growth so that forms of responsive organizational dynamism can be formed and developed. I will demonstrate these important synergies through three case studies that will show how the components of responsive organizational dynamism, strategic integration and cultural assimilation, actually operate in practice.

Siemens AG

The first case study offers a perspective from the chief information officer (CIO). The CIO of Siemens of the Americas at the time of this study was Dana Deasy, and his role was to introduce and expand the use of e-business across 20 discrete businesses. The Siemens Corporation worldwide network was composed of over 150 diverse sets of businesses, including transportation, healthcare, and telecommunications. Deasy's mission was to create a common road map across different businesses and cultures. What makes this case so distinct from others is that each business is highly decentralized under the umbrella of the Siemens Corporation. Furthermore, each company has its own mission; the companies have never been asked to come together and discuss common issues with regard to technology. That is, each business focused on itself as opposed to the entire organization. Deasy had to deal with two sectors of scope and hence, two levels of learning: the Americas as a region and the global firm internationally.

The challenge was to introduce a new e-business strategy from the top-down in each business in the Americas and then to integrate it with the global firm. Ultimately, the mission was to review what each business was doing in e-business and to determine whether there was an opportunity to consolidate efforts into a common direction.

IT was, for the most part, viewed as a back-office operation—handling services of the company as a support function as opposed to thinking about ways to drive business strategy. In terms of IT reporting, most CIOs reported directly to the chief financial officer (CFO). While some IT executives view this as a disadvantage because CFOs are typically too focused on financial issues, Deasy felt that a focus on cost containment was fine as long as the CIO had access to the chief executive officer (CEO) and others who ultimately drove business strategy. So, the real challenge was to ensure that CIOs had access to the various strategic boards that existed at Siemens.

What are the challenges in transforming an organization the size of Siemens? The most important issue was the need to educate CIOs on the importance of their role with respect to the business as opposed to the technology. As Deasy stated in an interview, "Business must come first and we need to remind our CIOs that all technology issues must refer back to the benefits it brings to the business." The question then is how to implement this kind of learning.

Perhaps the best way to understand how Siemens approached this dilemma is to understand Deasy's role as a corporate CIO. The reality is that there was no alternative but to create his position. What drove Siemens to this realization was fear that they needed someone to drive e-business, according to Deasy—fear of losing competitive edge in this area, fear that they were behind the competition and that smaller firms would begin to obtain more market share. Indeed, the growth of e-business occurred during the dot-com era, and there were huge pressures to respond to new business opportunities brought about by emerging technologies, specifically the Internet. It was, therefore, a lack of an internal capacity, such as responsive organizational dynamism, that stimulated the need for senior management to get involved and provide a catalyst for change.

The first aspect of Siemens's approach can be correlated to the strategic integration component of responsive organizational dynamism. We see that Siemens was concerned about whether technology was properly being integrated in strategic discussions. It established the Deasy role as a catalyst to begin determining the way technology needed to be incorporated within the strategic dimension of the business. This process cannot occur without executive assistance, so evolutionary learning must first be initiated by senior management. Unfortunately, Deasy realized early on that he needed a central process to allow over 25 CIOs in the Americas to interact regularly. This was important to understand the collective needs of the community and to pave the way for the joining of technology and strategic integration from a more global perspective. Deasy established an infrastructure to support open discourse by forming CIO forums, similar to communities of practice, in which CIOs came together to discuss common challenges, share strategies, and have workshops on the ways technology could help the business. Most important at these forums was the goal of consolidating their ideas and their common challenges.

There are numerous discussions regarding the common problems that organizations face regarding IT expenditures, specifically the approach to its valuation and return on investment (ROI). While there are a number of paper-related formulas that financial executives use (e.g., percentage of gross revenues within an industry), Deasy utilized learning theories, specifically, communities of practice, to foster more thinking and learning about what was valuable to Siemens, as opposed to using formulas that might not address important indirect benefits from technology. In effect, Deasy promoted learning among a relatively small but important group of CIOs who needed to better understand the importance of strategic innovation and the value it could bring to the overall business mission. Furthermore, these forums provided a place where CIOs could develop their own community—a community that allowed its members to openly participate in strategic discourse that could help transform the organization. It was also a place to understand the tacit knowledge of the CIO organization and to use the knowledge of the CIOs to summarize common practices and share them among the other members of the community.

Most of the CIOs at Siemens found it challenging to understand how their jobs were to be integrated into business strategy. Indeed, this is not a surprise. In Chapter 1, I discuss the feedback from my research on CEO evaluation of technology; I found that there were few IT executives who were actually involved in business strategy. Thus, the organization sought to create an advocate in terms of a centralized corporate headquarter that could provide assistance as opposed to forcing compliance. That is, it sought a structure with which to foster organizational learning concepts and develop an approach to create a more collective effort that would result in global direction for IT strategic integration.

To establish credibility among the CIO community, Deasy needed to ensure that the CIOs of each individual company were able to interact with board-level executives. In the case of Siemens, this board is called the president's council. The president's council has regularly held meetings in which each president attends and receives presentations on ideas about the regional businesses. Furthermore, there are quarterly CFO meetings as well, where CIOs can participate in understanding the financial implications of their IT investments. At the same time, these meetings provided the very exposure to the executive team that CIOs needed. Finally, Deasy established a CIO advisory board comprised of CIOs who actually vote on the common strategic issues and thus manage the overall direction of technology at Siemens. Each of these groups established different types of communities of practice that focused on a specific aspect of technology. The groups were geared to create better discourse and working relationships among these communities to, ultimately, improve Siemens's competitive advantage. The three communities of practice at work in the Siemens model— executive, finance, and technology—suggest that having only one general community of practice to address technology issues may be too limiting. Thus, theories related to communities of practice may need to be expanded to create discourse among multiple communities. This might be somewhat unique for IT, not in that there is a need for multiple communities, but that the same individuals must have an identity in each community. This shows the complexity of the CIO role today in the ability to articulate technology to different types and tiers of management. Figure 8.1 shows the interrelationships among the CIO communities of practice at Siemens.

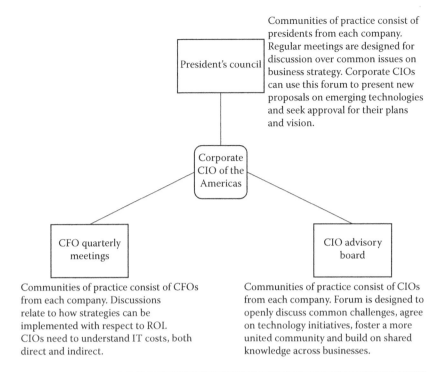

Figure 8.1   Inter-relationships among CIO communities of practice at Siemens.

Another way to represent these communities of practice is to view them as part of a process composed of three operating levels. Each level represents a different strategic role of management that is responsible for a unique component of discourse and on the authorization for uses of technology. Therefore, if the three different communities of practice are viewed strategically, each component could be constructed as a process leading to overall organizational cooperation, learning, and strategic integration as follows:

> *Tier 1: CIO Advisory Board*: This community discusses issues of technology standards, operations, communications, and initiatives that reflect technology-specific areas. Such issues are seen as CIO specific and only need this community's agreement and justification. However, issues or initiatives that require financial approval, such as those that may not yet be budgeted or approved, need to be discussed with group CFOs. Proposals to executive management—that is, the President's Council—also need prior approval from the CFOs.

Tier 2: CFO Quarterly: CFOs discuss new emerging technologies and ascertain their related costs and benefits (ROI). Those technologies that are already budgeted can be approved based on agreed ROI scenarios. Proposals for new technology projects are approved in terms of their financial viability and are prepared for further discussion at the President's Council.

Tier 3: President's Council: Proposals for new technology projects and initiatives are discussed with a focus on their strategic implications on the business and their expected outcome.

Deasy realized that he needed to create a common connection among these three communities. While he depended on the initiatives of others, he coordinated where these CIO initiatives needed to be presented, based on their area of responsibility.

Graphically, this can be shown as a linear progression of community-based discussions and approvals, as in Figure 8.2.

The common thread to all three tiers is the corporate CIO. Deasy was active in each community; however, his specific activities within each community of practice were different. CIOs needed to establish peer relationships with other CIOs share their tacit knowledge and contribute ideas that could be useful to other Siemens companies. Thus, CIOs needed to transform their personal views of technology and expand them to a group-level perspective. Their challenge was to learn how to share concepts and how to understand new ones that emanated at the CIO advisory board level. From this perspective, they could create the link between the local strategic issues and those discussed at the regional and global levels, as shown in Figure 8.3.

Using this infrastructure, Siemens's organizational learning in technology, occurred at two levels of knowledge management. The first is represented by Deasy's position, which effectively represents a top-down structure to initiate the learning process. Second, are the tiers of communities of practice when viewed hierarchically. This view reflects a more bottom-up learning strategy, with technological opportunities initiated by a community of regional, company CIOs, each representing the specific interests of their companies or specific lines of business. This view can also be structured as an evolutionary cycle in which top-down management is used to initiate organizational learning from the bottom-up, the bottom, in this case, represented by

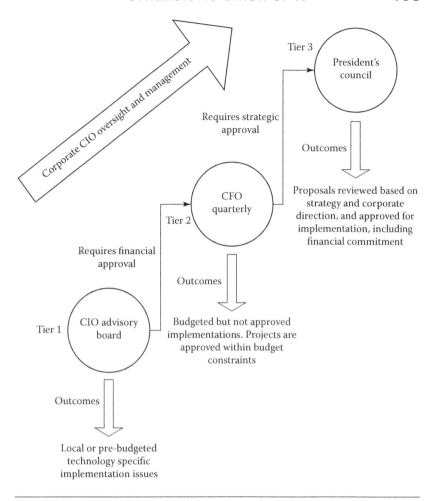

Corporate CIO oversight and management

Tier 3   President's council

Requires strategic approval

Outcomes

CFO quarterly

Tier 2

Proposals reviewed based on strategy and corporate direction, and approved for implementation, including financial commitment

Requires financial approval

Outcomes

Tier 1   CIO advisory board

Budgeted but not approved implementations. Projects are approved within budget constraints

Outcomes

Local or pre-budgeted technology specific implementation issues

Figure 8.2   Siemens' community-based links.

local operating company CIOs. This means that the CIO is seen relatively, in this case, as the lower of the senior management population. Figure 8.4 depicts the CIO as this "senior lower level."

From this frame of reference, the CIO represents the bottom-up approach to the support of organizational learning by addressing the technology dilemma created by technological dynamism—specifically, in this case, e-business strategy.

The role of IT in marketing and e-business was another important factor in Siemens's model of organizational learning. The technology strategy at Siemens was consistent with the overall objectives of the organization: to create a shared environment that complements each

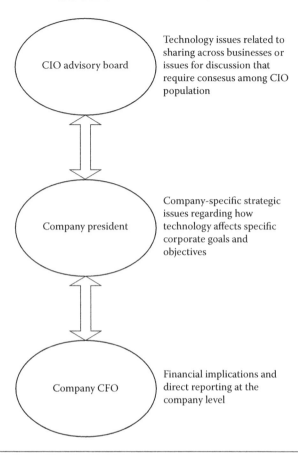

Technology issues related to sharing across businesses or issues for discussion that require consesus among CIO population

CIO advisory board

Company-specific strategic issues regarding how technology affects specific corporate goals and objectives

Company president

Financial implications and direct reporting at the company level

Company CFO

Figure 8.3 Siemens' local to global links.

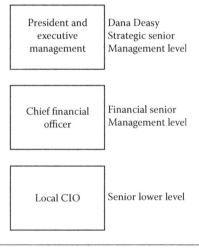

President and executive management — Dana Deasy Strategic senior Management level

Chief financial officer — Financial senior Management level

Local CIO — Senior lower level

Figure 8.4 CIO as the "senior lower level."

business by creating the opportunity to utilize resources. This shared environment became an opportunity for IT to lead the process and become the main catalyst for change. I discuss this kind of support in Chapter 5, in which I note that workers see technology as an acceptable agent of change. Essentially, the CIOs were challenged with the responsibility of rebranding their assets into clusters based on their generic business areas, such as hospitals, medical interests, and communications. The essence of this strategic driver was to use e-business strategy to provide multiple offerings to the same customer base.

As with the Ravell case discussed in Chapter 1, the Siemens case represents an organization that was attempting to identify the driver component of IT. To create the driver component, it became necessary for executive management to establish a corporate position (embodied by Deasy) to lay out a plan for transformation, through learning and through the use of many of the organizational learning theories presented in Chapter 4.

The Siemens challenge, then, was to transform its CIOs from being back-office professionals to proactive technologists focused primarily on learning to drive business strategy. That is not to say that back-office issues became less important; they became, instead, responsibilities left to the internal organizations of the local CIOs. However, back-office issues can often become strategic problems, such as with the use of e-mail. This is an example of a driver situation even though it still pertains to a support concern. That is, back-office technologies can indeed be drivers, especially when new or emerging technologies are available. As with any transition, the transformation of the CIO role was not accomplished without difficulty. The ultimate message from executive management to the CIO community was that it should fuse the vital goals of the business with its technology initiatives. Siemens asked its CIOs to think of new ways that technology could be used to drive strategic innovations. It also required CIOs to change their behavior by asking them to think more about business strategy.

The first decision that Deasy confronted was whether to change the reporting structure of the CIO. Most CIOs at Siemens reported directly to the CFO as opposed to the CEO. After careful thought, Deasy felt that to whom the CIO reported was less important than giving access and exposure to the President's Council meetings. It was Deasy's perspective that only through exposure and experience could

CIOs be able to transform from back-office managers to strategic planners. As such, CIO training was necessary to prepare them for participation in communities of practice. Eventually, Siemens recognized this need and, as a result, sponsored programs, usually lasting one week, in which CIOs would be introduced to new thinking and learning by using individual-based reflective practices. Thus, we see an evolutionary approach, similar to that of the responsive organizational dynamism arc, presented in Chapter 4; that is, one that uses both individual and organizational learning techniques.

Deasy also understood the importance of his relationship and role with each of the three communities of practice. With respect to the CEOs of each company, Deasy certainly had the freedom to pick up the phone and speak with them directly. However, this was rarely a realistic option as Deasy knew early on that he needed the trust and cooperation of the local CIO to be successful. The community with CEOs was then broadened to include CIOs and other senior managers. This was another way in which Deasy facilitated the interaction and exposure of his CIOs to the executives at Siemens.

Disagreement among the communities can and does occur. Deasy believed in the "pushing-back" approach. This means that, inevitably, not everyone will agree to agree, and, at times, senior executives may need to press on important strategic issues even though they are not mutually in agreement with the community. However, while this type of decision appears to be contrary to the process of learning embedded in communities of practice learning, it can be a productive and acceptable part of the process. Therefore, while a democratic process of learning is supported and preferred, someone in the CIO position ultimately may need to make a decision when a community is deadlocked.

The most important component of executive decision making is that trust exists within the community. In an organizational learning infrastructure, it is vital that senior management share in the value proposition of learning with members of the community. In this way, members feel that they are involved, and are a part of decision making as opposed to feeling that they are a part of a token effort that allows some level of participation. As Deasy stated, "I was not trying to create a corporate bureaucracy, but rather always representing myself as an ambassador for their interest, however, this does not guarantee that I will always agree with them." Disagreements, when managed properly, require patience, which can result in iterative discussions with members of the community before a consensus position may be reached, if it is at all. Only after this iterative process is exhausted does a senior overarching decision need to be made. Deasy attributed his success to his experience in field operations, similar to those of his constituents. As a prior business-line CIO, he understood the dilemma that many members of the community were facing. Interestingly, because of his background, Deasy was able to "qualify" as a true member of the CIO community of practice. This truth establishes an important part of knowledge management and change management—senior managers who attempt to create communities of practice will be more effective when they share a similar background and history with the community that they hope to manage. Furthermore, leaders of such communities must allow members to act independently and not confuse that independence with autonomy. Finally, managers of communities of practice are really champions of their group and as such must ensure that the trust among members remains strong. This suggests that CIO communities must first undergo their own cultural assimilation to be prepared to integrate with larger communities within the organization.

Another important part of Deasy's role was managing the technology itself. This part of his job required strategic integration in that his focus was more about uses of technology, as opposed to community behavior or cultural assimilation. Another way of looking at this issue is to consider the ways in which communities of practice actually transform tacit knowledge and present it to senior management as explicit knowledge. This explicit knowledge about uses of technology must be presented in a strategic way and show the benefits for the organization. The ways that technology can benefit a business often reside within IT as tacit knowledge. Indeed, many senior managers often criticize IT managers for their inability to articulate what they know and to describe it so that managers can understand what it means to the business. Thus, IT managers need to practice transforming their tacit knowledge about technology and presenting it effectively, as it relates to business strategy.

Attempting to keep up with technology can be a daunting, if not impossible, task. In some cases, Siemens allows outside consultants to provide help on specific applications if there is not enough expertise within the organization. The biggest challenge, however, is not necessarily in keeping up with new technologies but rather, in testing technologies to determine exactly the benefit they have on the business. To address this dilemma, Deasy established the concept of "revalidation." Specifically, approved technology projects are reviewed every 90 days to determine whether they are indeed providing the planned outcomes, whether new outcomes need to be established, or whether the technology is no longer useful. The concept of revalidation can be associated with my discussion in Chapter 3, which introduced the concept of "driver" aspects of technology. This required that IT be given the ability to invest and experiment with technology to fully maximize the evaluation of IT in strategic integration. This was particularly useful to Deasy, who needed to transform the culture at Siemens to one that recognized that not all approved technologies succeed. In addition, he needed to dramatically alter the application development life cycle and reengineer the process of how technology was evaluated by IT and senior management. This challenge was significant in that it had to be accepted by over 25 autonomous presidents, who were more focused on short and precise outcomes from technology investments.

Deasy was able to address the challenges that many presidents had in understanding IT jargon, specifically as it related to benefits of using technology. He engaged in an initiative to communicate with non-IT executives by using a process called *storyboarding*. Storyboarding is the process of creating prototypes that allow users to actually see examples of technology and how it will look and operate. Storyboarding tells a story and can quickly educate executives without being intimidating. Deasy's process of revaluation had its own unique life cycle at Siemens:

1. Create excitement through animation. What would Siemens be like if ... ?
2. Evaluate the way the technology would be supported.
3. Recognize implementation considerations about how the technology as a business driver is consistent with what the organization is doing and experiencing.

4. Technology is reviewed every 90 days by the CIO advisory board after experimental use with customers and presented to the president's council on an as-needed basis.

5. Establish responsive organizational dynamism with cultural assimilation; that is, recognize the instability of technology and that there are no guarantees to planned outcomes. Instead, promote business units to understand the concept of "forever prototyping."

Thus, Siemens was faced with the challenge of cultural assimilation, which required dramatic changes in thinking and business life cycles. This process resembles Bradley and Nolan's (1998) *Sense and Respond*—the ongoing sensing of technology opportunities and responding to them dynamically. This process disturbs traditional and existing organizational value chains and therefore represents the need for a cultural shift in thinking and doing. Deasy, using technology as the change variable, began the process of reinventing the operation of many traditional value chains.

Siemens provides us with an interesting case study for responsive organizational dynamism because it had so many diverse companies (in over 190 countries) and over 425,000 employees. As such, Siemens represents an excellent structure to examine the importance of cultural assimilation. Deasy, as a corporate CIO, had a counterpart in Asia/Australia. Both corporate CIOs reported to a global CIO in Germany, the home office of Siemens. There was also a topic-centered CIO responsible for global security and application-specific planning software. This position also reported directly to the global CIO. There were regional and local CIOs who focused on specific geographical areas and vertical lines of business and operating company CIOs. This organization is shown in Figure 8.5.

Deasy's operation represents one portion (although the most quickly changing and growing) of Siemens worldwide. Thus, the issue of globalization is critical for technologies that are scalable beyond regional operating domains. Standardization and evaluations of technology often need to be ascertained at the global level and as a result introduce new complexities relating to cultural differences in business methods and general thinking processes. Specifically, what works in one country may not work the same way in another. Some of these

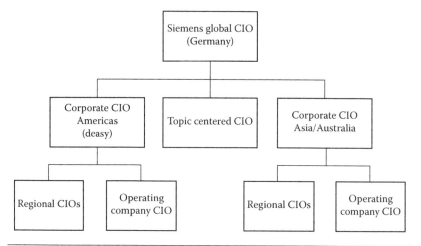

Figure 8.5 Siemens' CIO organization.

matters can be legally based (e.g., licensing of software or assumptions about whether a technology is legally justified). To a large extent, solving legal matters relating to technology is easier than cultural ones.

Cultural assimilation matters about technology typically occur in global organizations with respect to acceptability of operational norms from one country to another. This becomes a particularly difficult situation when international firms attempt to justify standards. At Siemens, Deasy introduced three "standards" of technology that defined how it could be used across cultures, and communities of practice:

1. *Corporate services*: These are technologies that are required to be used by the business units. There are central service charges for their use as well.
2. *Mandatory services*: Everyone must comply with using a particular type of application; that is, mandatory software based on a specific type of application. For example, if you use a Web browser, it must be Internet Explorer.
3. *Optional*: These are technologies related to a specific business and used only within a local domain. There may be a preferred solution, but IT is not required to use it.

This matrix of standards allows for a culture to utilize technologies that are specific to its business needs, when justified. Standards at Siemens are determined by a series of steering committees, starting at the regional level, that meet two to three times annually. Without question, implementing standards across cultures is, as Deasy phrased it, "a constant wrestling match which might need to change by the time a standard is actually reached." This is why strategic integration is so important, given the reality that technology cannot always be controlled or determined at senior levels. Organizations must be able to dynamically integrate technology changes parallel to business changes.

Deasy's longer-term mission was to provide a community of CIOs who could combine the business and technology challenges. It was his initial vision that the CIO of the future would be more involved than before with marketing and value chain creation. He felt that "the CIO community needed to be detached from its technology-specific issues or they would never be a credible business partner." It was his intent to establish organizational learning initiatives that helped CIOs "seize and succeed," to essentially help senior management by creating vision and excitement, by establishing best practices, and by learning better ways to communicate through open discourse in communities of practice.

Three years after his initial work, I reviewed the progress that Deasy had made at Siemens. Interestingly, most of his initiatives had been implemented and were maturing—except for the role of e-business strategy. I discovered, after this period, that the organization thought that e-business was an IT responsibility. As such, they expected that the CIOs had not been able to determine the best business strategy. This was a mistake; the CIO could not establish strategy but rather needed to react to the strategies set forth by senior management. This means that the CIO was not able to really establish stand-alone strategies as drivers based on technology alone. CIOs needed, as Deasy stated, "to be a participant with the business strategist and to replace this was inappropriate." This raises a number of questions:

1. Did this occur because CIOs at Siemens do not have the education and skills to drive aspects of business strategy?
2. Did the change in economy and the downfall of the dot-coms create a negative feeling toward technology as a business driver?

3. Are CEOs not cognizant enough about uses of technology, and do they need better education and skills to better understand the role of technology?
4. Is the number of communities of practice across the organization integrated enough so that IT can effectively communicate and form new cultures that can adapt to the changes brought about by emerging technologies?
5. Is there too much impatience with the evolution of technology? Does its assimilation in an organization the size of Siemens simply take too long to appreciate and realize the returns from investments in technology?

I believe that all of these questions apply, to some extent, and are part of the challenges that lie ahead at Siemens. The company has now initiated a series of educational seminars designed to provide more business training for CIOs, which further emphasizes the importance of focusing on business strategy as opposed to just technology. It could also mean the eventual establishment of a new "breed" of CIOs who are better educated in business strategy. However, it is inappropriate for non-IT managers to expect that the CIOs will be able to handle strategy by themselves; they must disconnect e-business as solely being about technology. The results at Siemens only serve to strengthen the concept that responsive organizational dynamism requires that cultural assimilation occur within all the entities of a company.

Aftermath

Dana Deasy left Siemens a few years after this case study was completed. During that time, the executive team at Siemens realized that the CIO alone could not provide business strategy or react quickly enough to market needs. Rather, such strategy required the integration of all aspects of the organization, with the CIO only one part of the team to determine strategic shifts that lead or use components of technology. Thus, the executives realized that they needed to become much better versed in technology so that they also could engage in strategic conversations. This does not suggest that executives needed technology training per se, but that they do need training that allows them to comment intelligently on technology issues. What is the best way to accomplish this goal? The answer is through short seminars that can provide executives with terminology and familiarize them with the processes their decisions will affect. The case also raised the question of whether a new wave of executives would inevitably be required to move the organization forward to compete more effectively. While these initiatives appear to make sense, they still need to address the fundamental challenges posed by technology dynamism and the need to develop an organization that is positioned to respond (i.e., responsive organizational dynamism). We know from the results of the Ravell case that executives cannot be excluded. However, the case also showed that all levels of the organization need to be involved. Therefore, the move to responsive organizational dynamism requires a reinvention of the way individuals work, think, and operate across multiple tiers of management and organizational business units. This challenge will continue to be a difficult but achievable objective of large multinational companies.

ICAP

This second case study focuses on a financial organization called ICAP, a leading money and securities broker. When software development exceeded 40% of IT activities, ICAP knew it was time to recognize IT as more than just technical support. Stephen McDermott provided the leadership, leaving his role as CEO of the Americas at ICAP to become CEO of the Electronic Trading Community (ETC), a new entity focused solely on software development. This IT community needed to be integrated with a traditional business model that was undergoing significant change due to emerging technologies, in this specific case, the movement from voice to electronic trading systems.

This case study reflects many aspects of the operation of responsive organizational dynamism. From the strategic integration perspective, ICAP needed to understand the ways electronic trading could ultimately affect business strategy. For example, would it replace all voice-related business interactions, specifically voice trading? Second, what would be the effect on its culture, particularly with respect to the way the business needed to be organizationally structured? This study focuses on the role of the CEO as a pioneer in reexamining his own biases, which favored an old-line business process, and for developing a realization to manage a major change in business strategy and organizational philosophy. Indeed, as McDermott stated, "It was the challenge of operating at the top, yet learning from the bottom." This sentiment essentially reflects the reality of a management dilemma. Could a CEO who, without question, had substantial knowledge of securities trading, learn to lead a technology-driven operation, for which he had little knowledge and experience?

To better understand the impact of technology on the business of ICAP, it is important to have some background information. Since 1975, the use of technology at ICAP was limited to operations of the back-office type. Brokers (the front-end or sales force of a trading business), communicated with customers via telephone. As such, processing transactions was always limited to the time necessary to manually disseminate prices and trading activity over the phone to a securities trader. However, by 1997 a number of technological advancements, particularly with the proliferation of Internet-based communication and the increased bandwidth available, enabled brokers and dealers to communicate bidirectionally. The result was that every aspect of the trade process could now be streamlined, including the ability for the trader to enter orders directly into the brokers' trading systems. The technological advancements and the availability of capital in the mid-1990s made it difficult to invest in computer operations. Specifically, the barriers to investing in technology had been high as developing proprietary trading systems and deploying a private network were all costly. The market of available products was scarce, filled with relatively tiny competitors with little more than a concept, rather than an integrated product that could do what a company like ICAP needed, in order to maintain its competitive position. The existing system, called the ICAP Trading Network application was far from a trading system that would compete against the newer emerging technologies. The goal was to develop a new trading system that would establish an electronic link between the back-office systems of ICAP and its clients. The system would need to be simple to use as the traders were not necessarily technology literate. It would need to be robust, include features that were specific to the markets, and easily installed and distributed. In addition, as ICAP decided to fund the entire project, it would have to be cost-effective and not burden the other areas of the business. As competitive systems were already being introduced, the new system needed to be operational within three to six months for ICAP to remain competitive.

McDermott recognized that designing a new product would require that IT developers and business matter experts learn to work together. As a result of this realization, a representative from the operation was selected to see if a third-party developer could modify an existing product. After exploring and evaluating responses, the search team concluded that off-the-shelf solutions, prohibitive in cost, were not available that would meet the critical timing needs of the business. However, during the period when IT and the business users worked together, these groups came to realize that the core components of its own trading system could be modified and used to build the new system. This realization resulted from discussions between IT and the business users that promoted organizational learning. This process resembles the situation in the Ravell study, in which I concluded that specific events could accelerate organizational learning and actually provide an opportunity to embed the process in the normal discourse of an organization. I also concluded that such learning starts with individual reflective practices, and understanding how both factions, in this case, IT and the business community, can help each other in a common cause. In the case of Ravell, it was an important relocation of the business that promoted integration between IT and the business community. At ICAP, the common cause was about maintaining competitive advantage.

The project to develop the new electronic trading application was approved in August 1999, and the ETC was formed. The new entity included an IT staff and selected members from the business community, who moved over to the new group. Thus, because of technological dynamism, it was determined that the creation of a new product established the need for a new business entity that would form its own strategic integration and cultural assimilation. An initial test of the new product took place in November, and it successfully executed the first electronic trade via the Internet. In addition to their design responsibility, ETC was responsible for marketing, installing, and training clients on the use of the product. The product went live in February 2000. Since its introduction, the ETC product has been modified to accommodate 59 different fixed-income products, serving more than 1,000 users worldwide in multiple languages.

While the software launch was successful, McDermott's role was a challenge, from coordinating the short- and long-term goals of ETC with the traditional business models of ICAP to shifting from management of a global financial enterprise to management of an IT community. The ICAP case study examines the experiences and perceptions one year after the launch of the new entity.

The first most daunting result, after a year of operations, was the significant growth of technology uses in the business. Initially, McDermott noted that electronic trading was about 40% of operations and that it had grown over 60%. He stated that ETC had become, without question, the single most important component of the ICAP international business focus. The growth of electronic trading created an accelerated need for transformation within ICAP and its related businesses. This transformation essentially changed the balance between voice or traditional trading and electronic trading. McDermott found himself responsible for much of this transformation and was initially concerned whether he had the technical expertise to manage it.

McDermott admitted that as a chief executive of the traditional ICAP business, he was conservative and questioned the practicality and value of many IT investments. He often turned down requests for more funding and looked at technology as more of a supporter of the business. As I explain in Chapter 3, IT as a supporter will always be managed, based on efficiencies and cost controls. McDermott's view was consistent with this position. In many ways, it was ironic that he became the CEO of the electronic component of the business. Like many CEOs, McDermott initially had the wrong impression of the Internet. Originally looking at it as a "big threat," he eventually realized from the experience that the Internet was just another way of communicating with his clients and that its largest contribution was that it could be done more cost-effectively, thus leading to higher profits.

One of the more difficult challenges for McDermott was developing the mission for ETC. At the time of the launch of the new product, this mission was unclear. With the assistance of IT and the business community, the mission of ETC has been developing dynamically; the business is first trying to protect itself from outside competition. Companies like IBM, Microsoft, and others, might attempt to invade the business market of ICAP. Thus, it is important that ETC continues to produce a quality product and keep its competitive edge over more limited competitors that are software-based organizations only. The concept of a dynamic mission can be correlated to the fundamental principles of responsive organizational dynamism. In fact, it seems rather obvious that organizations dealing with emerging technologies might need to modify their missions to parallel the accelerated changes brought about by technological innovation. We certainly see this case with ICAP, for which the market conditions became volatile because of emerging electronic trading capacities. Why, then, is it so difficult for organizations to realize that changing or modifying their missions should not be considered that unusual? Perhaps the approach of ICAP in starting a completely separate entity was correct. However, it is interesting that this new organization was operating without a consistent and concrete mission.

Another important concept that developed at ETC was that technology was more of a commodity and that content (i.e., the different services offered to clientele) was more important. Indeed, as McDermott often stated, "I assume that the technology works, the real issue is the way you intend to implement it; I want to see a company's business plan first." Furthermore, ETC began to understand that technology could be used to leverage ICAP businesses in areas that they had never been able to consider before the advent of the technology and the new product. McDermott knew that this was a time, as Deasy often stated, to "seize and succeed" the moment. McDermott also realized that organizational learning practices were critical for ideas to come from within the staff. He was careful not to require staff to immediately present a formal new initiative, but he allowed them to naturally develop a plan as the process became mature. That is one of the reasons that ETC uses the word *community* in its name. As he expressed it to me during a conversation:

Now that is not my mandate to grow into other areas of opportunity, my initial responsibility is always to protect our businesses. However, I will not let opportunities go by which can help the business grow, especially things that we could never do as a voice broker. It has been very exciting and I can see ICAP becoming a considerably larger company than we have been historically because of our investment in technology.

McDermott also was challenged to learn what his role would be as a chief executive of a software technology organization. In the early stages, he was insecure about his job because for the first time he knew less than his workers about the business. Perhaps this provides organizational learning practitioners with guidance on the best way of getting the CEO engaged in the transformative process; that is, getting the CEO to understand his or her role in an area in which, typically, he or she does not have expertise. McDermott represented an executive who reached that position coming up through the ranks. Therefore, much of his day-to-day management was based on his knowledge of the business—a business that he felt he knew as well as anyone. With technology, and its effect as technological dynamism, CEOs face more challenges, not only because they need to manage an area they may know little about but also because of the dynamic aspects of technology and the way it causes unpredictable and accelerated change. McDermott realized this and focused his attention on discovering what his role needed to be in this new business. There was no question in McDermott's mind that he needed to know more about technology, although he also recognized that management was the fundamental responsibility he would have with this new entity:

> [Although] I was insecure at the beginning I started to realize that it does not take a genius to do my job. Management is management, and whether you manage a securities brokering firm or you manage a deli or manage a group of supermarkets or an IT or an electronic company, it is really about management, and that is what I am finding out now. So, whether I am the right person to bring ETC to the next level is irrelevant at this time. What is more important is that I have the skills that are necessary to manage the business issues as opposed to the technological ones.

However, McDermott did have to make some significant changes to operate in a technology-based environment. ETC was now destined to become a global organization. As a result, McDermott had to create three senior executive positions to manage each of the three major geographic areas of operation: North America, Europe, and Asia. He went from having many indirect reports to having just a few. He needed four or five key managers. He needed to learn to trust that they were the right people, people who had the ability to nurture the parts of each of their respective divisions. "What it leaves now is being a true CEO," he stated, "and that means picking your people, delegating the responsibility and accepting that they know the business." Thus, we see technological dynamism actually realigning the reporting structure, and social discourse of the company.

My presentation in previous chapters focused on helping organizations transform and change. Most important in organizational learning theories is the resistance to change that most workers have, particularly when existing cultural norms are threatened. ICAP was no exception to the challenges of change management. The most significant threat at ICAP was the fear that the traditional voice broker was endangered. McDermott understood this fear factor and presented electronic trading not as a replacement but rather, a supplement to the voice broker. There was no question that there were certain areas of the business that lent themselves more to electronic trading; however, there are others that will never go electronic or at least predominantly electronic. Principles of responsive organizational dynamism suggest that accelerated change becomes part of the strategic and cultural structure of an organization. We see both of these components at work in this case.

Strategically, ICAP was faced with a surge in business opportunities that were happening at an accelerated pace and were, for the most part, unplanned, so there was little planned activity. The business was feeling its way through its own development, and its CEO was providing management guidance, as opposed to specific solutions. ICAP represents a high-velocity organization similar to those researched by Eisenhardt and Bourgeois (1988), and supports their findings that a democratic, less power-centralized management structure enhances the performance of such a firm. From a cultural assimilation perspective, the strategic decisions are changing the culture and requiring new structures and alignments. Such changes are bound to cause fears.

As a result of recognizing the inevitable changes that were becoming realities, McDermott reviewed the roles and responsibilities of his employees on the brokering side of the business. After careful analysis, he realized that he could divide the brokers into three different divisions, which he branded as A, B, and C brokers. The A brokers were those who were fixed on the relational aspect of their jobs, so voice interaction was the only part of their work world. Such individuals could do things in the voice world that electronic means could not reach. They were personal experts, if you will, who could deal with clients requiring a human voice. Thus, the A broker would exist as long as the broker wanted to work—and would always be needed because a population of clients wants personal support over the phone. This is similar to the opposition to the Internet in which we find that some portion of the population will never use e-commerce because they prefer a live person. The B broker was called the hybrid broker—an individual who could use both voice and electronic means. Most important, these brokers were used to "convert" voice-based clients into electronic ones. As McDermott explained:

> Every day I see a different electronic system that someone is trying to sell in the marketplace. Some of these new technologies are attempting to solve problems that do not exist. I have found that successful systems address the content more than the technology. Having a relationship for many of our customers is more important. And we can migrate those relationships from voice to electronic or some sort of a hybrid combination. The B brokers will end up with servicing some combination of these relationships or migrate themselves to the electronic system. So, I believe they have nothing to fear.

The C brokers, on the other hand, represented the more average voice brokers who would probably not have a future within the business. They would be replaced by electronic trading because they did not bring the personal specialization of the A broker. The plight of the C broker did raise an important issue about change management and technological dynamism: Change will cause disruption, which can lead to the elimination of jobs. This only further supported the fears that workers had when faced with dynamic environments. For McDermott, this change would need to be openly discussed with the community, especially for the A and B brokers, who in essence would continue to play an important role in the future of the business. C brokers needed to be counseled so that they could appropriately seek alternate career plans. Thus, honesty brings forth trust, which inevitably fosters the growth of organizational learning. Another perspective was that the A and B brokers understood the need for change and recognized that not everyone could adapt to new cultures driven by strategic integration, so they understood why the C broker was eliminated.

In Chapter 2, I discussed the dilemma of IT as a "marginalized" component of an organization. This case study provides an opportunity to understand how the traditional IT staff at ICAP made the transition into the new company—a company in which they represented a direct part of its success. As noted, ICAP considered the IT department as a back-office support function. In the new organization, it represented the nucleus or the base of all products and careers. Hence, McDermott expected ETC employees to be technology proficient. No longer were IT people just coders or hardware specialists—he saw technology people as lawyers, traders, and other businesspeople. He related technology proficiency in a similar way to how his business used to view a master's degree in business (MBA) in the late 1980s. This issue provides further support for the cultural assimilation component of responsive organizational dynamism. We see a situation in which the discrepancy between who is and is not a technology person beginning to dwindle in importance. While there is still clear need for expertise and specialization, the organization as a whole has started the process of educating itself on the ways in which technology affects every aspect of its corporate mission, operations, and career development.

ICAP has not been immune to the challenges that have faced most technology-driven organizations. As discussed in Chapter 2, IT projects typically face many problems in terms of their ability to complete projects on time and within budget. ICAP was also challenged with this dilemma. Indeed, ICAP had no formal process but focused on the criterion of meeting the delivery date as the single most important issue. As a result, McDermott was attempting to instill a new culture committed to the importance of what he called the "real date of delivery." It was a challenge to change an existing culture that had difficulty with providing accurate dates for delivery. As McDermott suggested:

> I am learning that technology people know that there is no way that they can deliver an order in the time requested, but they do not want to disappoint us. I find that technology people are a different breed from the people that I normally work with. Brokers are people looking for immediate gratification and satisfaction. Technology people, on the other hand, are always dedicated to the project regardless of its time commitment.

McDermott was striving to attain a mix or blend of the traditional culture with the technology culture and create a new hybrid organization capable of developing realistic goals and target dates. This process of attainment mirrors the results from the Ravell case, which resulted in the formation of a new hybrid culture after IT and business staff members were able to assimilate one another and find common needs and uses for technology and the business.

McDermott also understood his role as a leader in the new organization. He realized early on that technology people are what he called more "individualistic"; that is, they seemingly were reluctant to take on responsibility of other people. They seemed, as McDermott observed, "to have greater pleasure in designing and creating something and they love solving problems." This was different from what CEOs experienced with MBAs, who were taught more to lead a group as opposed to being taught to solve specific problems. Yet, the integration of both approaches can lead to important accomplishments that may not be reachable while IT and non-IT are separated by departmental barriers.

Ultimately, the cultural differences and the way they are managed lead to issues surrounding the basis of judging new technologies for future marketing consideration. McDermott understood that this was a work in progress. He felt strongly that the issue was not technology, but that it was the plan for using technology competitively. In other words, McDermott was interested in the business model for the technology that defined its benefits to the business strategically. As he put it, "Tell me how you are going to make money, tell me what you can do for me to make my life easier. That is what I am looking at!" While McDermott felt that many people were surprised by his response, he believed its reality was taken too much for granted. During the dot-com era, too many investors and businesses assumed that technological innovation would somehow lead to multiples of earnings—that simply did not happen. Essentially, McDermott realized that good technology was available in many places and that the best technology is not necessarily the one that will provide businesses with the highest levels of success.

Judging new technologies based on the quality of the business plan is an effective method of emphasizing the importance of why the entire organization needs to participate and understand technology.

This inevitably leads to questions about the method in which ROI is, or should be, measured. The actual measurement of ROI for ICAP was remarkably simple yet effective. There were four methods of determining ROI. The first and most significant was whether the technology would increase volume of trades along the different product lines. The second was the amount in dollars of the securities being traded. That is, did technology provide a means for clients to do larger dollar trades? The third factor could be an increase in the actual number of clients using the electronic system. The fourth might be alleviating existing bottlenecks in the voice trading process, whether it was a legal issue or the advantage provided by having electronic means. We see here that some of the ROI factors are direct and monetary. As expected methods, the first and second were very much direct monetary ways to see the return for investing in electronic trading systems. However, as Lucas (1999) reminds us, many benefits derived from IT investments are indirect, and some are impossible to measure. We see this with the third and fourth methods. Increasing the number of clients indirectly suggested more revenue, but did not guarantee it. An even more abstract benefit was the improvement of throughput, what is typically known as improved efficiency in operations.

While all of the accomplishments of ICAP and McDermott seem straightforward, they were not accomplished without challenges; perhaps the most significant was the approach, determination, and commitment that were needed by the executive team. This challenge is often neglected in the literature on organizational learning. Specifically, the executive board of ETC needed to understand what was necessary in terms of funding to appropriately invest in the future of technology. To do that, they needed to comprehend what e-business was about and why it was important for a global business to make serious investments in it to survive. In this context, then, the executive board needed to learn about technology as well and found themselves in a rather difficult position. During this period, McDermott called in an outside consultant who could provide a neutral and objective opinion. Most important was to define the issue in lay terms so that board members could correlate it with their traditional business models. Ultimately, the learning consisted of understanding that technology and e-commerce were about expanding into more markets, ones that ICAP could not reach using traditional approaches. There was a realization that ICAP was too focused on its existing client base, as opposed to reaching out for new ones—and there was also the reverse reality that a competitor would figure out a strategy to reach out to the client base of ICAP. What is also implied in expanding one's client base is that it means going outside one's existing product offerings. This had to be carefully planned as ICAP did not want to venture outside what it was—an intermediary brokering service. So, expansion needed to be carefully planned and discussed first among the executive members, then presented as a challenge to the senior management, and so on.

This process required some modifications to the organizational learning process proposed by such scholars as Nonaka and Takeuchi (1995). Specifically, their models of knowledge management do not typically include the executive boards; thus, they are not considered a part of the learning organization. The ICAP case study exposes the fact that their exclusion can be a serious limitation, especially with respect to the creation of responsive organizational dynamism. In previous chapters, I presented a number of management models that could be used to assist in developing and sustaining organizational learning. They focused fundamentally on the concept of whether such management should be top-down, bottom-up, or, as Nonaka and Takeuchi suggest, "middle-up-down." I laid out my case for a combination of all of them in a specific order and process that could maximize each approach. However, none of these models really incorporates the outside executive boards that have been challenged to truly understand what technology is about, their approach to management, and what their overall participation should be in organizational learning.

Perhaps the most significant historical involvement of executive boards was with the Year 2000 (Y2K) event. With this event, executive boards mandated that their organizations address the potential technology crisis at the turn of the century. My CEO interviews verified that, if anything, the Y2K crisis served to educate executive boards by forcing them to focus on the issue. Boards became unusually involved with the rest of the organization because independent accounting firms, as outside objective consultants, were able to expose the risks for not addressing the problem. The handling of e-commerce by ICAP was in many ways similar but also suggests that executive boards should not always wait for a crisis to occur before they get involved. They also must be an important component of organizational learning, particularly in responsive organizational dynamism. While organizational learning fosters the involvement of the entire community or workers, it also needs advocates and supporters who control funding. In the case of ICAP, organizational learning processes without the participation of the executive board, ultimately would not have been successful. The experience of ICAP also suggests that this educational and learning process may need to come from independent and objective sources, which integrates another component of organizational learning that has not been effectively addressed: the role of outside consultants as a part of a community of practice. Figure 8.6 depicts the addition of the ICAP ETC executive board and outside consultants in the organizational learning management process.

The sequential activities that occurred among the different communities are shown in Table 8.1. While Table 8.1 shows the sequential steps necessary to complete a transformation toward strategic integration and cultural assimilation, the process is also very iterative. Specifically, this means that organizations do not seamlessly move from one stage to another without setbacks. Thus, transformation depends heavily on discourse as the main driver for ultimate organizational evolution.

Figure 8.7 shows a somewhat messier depiction of organizational learning under the auspices of ROD. The changes brought on by dynamic interactions foster top-down, middle-up-down, and bottom-up knowledge management techniques—all occurring simultaneously. This level of complex discourse creates a number of overlapping communities of practice that have similar, yet unique, objectives in learning. These communities of practice overlap at certain levels as shown in Figure 8.8.

As stated, organizational learning at the executive levels tends to be ignored in the literature. At ICAP, an important community of practice emerged that created a language discourse essential to its overall success in dealing with technological dynamism, brought on by technological innovation in electronic communications. Language was critical at this level; ICAP is a U.K.-based organization and as such has an international board. As McDermott explained:

> As you know, from travelling anywhere around the world, cultures are
> different. And even the main office for our company, ICAP in England,

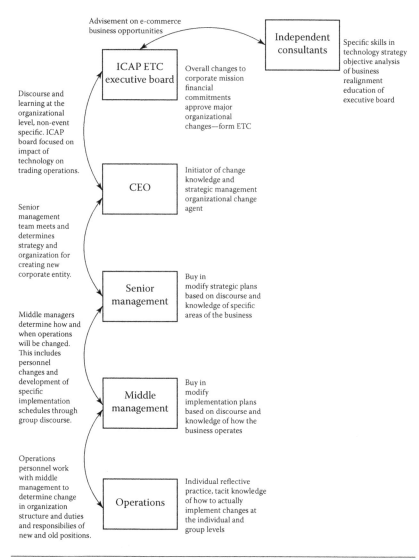

Advisement on e-commerce
business opportunities

Independent
consultants

Specific skills in
technology strategy
objective analysis
of business
realignment
education of
executive board

ICAP ETC
executive board

Overall changes to
corporate mission
financial
commitments
approve major
organizational
changes—form ETC

Discourse and
learning at the
organizational
level, non-event
specific. ICAP
board focused on
impact of
technology on
trading operations.

CEO

Initiator of change
knowledge and
strategic management
organizational change
agent

Senior
management
team meets and
determines
strategy and
organization for
creating new
corporate entity.

Senior
management

Buy in
modify strategic plans
based on discourse and
knowledge of specific
areas of the business

Middle managers
determine how and
when operations
will be changed.
This includes
personnel
changes and
development of
specific
implementation
schedules through
group discourse.

Middle
management

Buy in
modify
implementation plans
based on discourse and
knowledge of how the
business operates

Operations
personnel work
with middle
management to
determine change
in organization
structure and duties
and responsibilities of
new and old positions.

Operations

Individual reflective
practice, tacit knowledge
of how to actually
implement changes at
the individual and
group levels

Figure 8.6 ICAP ETC management tiers.

and even with the English, we are separated by a common language, as we often say. There is a very, very different culture everywhere in the world. I will tell you that information technology in our company is separated from electronic trading—there is a difference.

Thus, McDermott's challenge was to establish a community that could reach consensus not only on strategic issues but also

Table 8.1   ICAP—Steps to Transformation

STEP	LEARNING ENTITY(S)	LEARNING ACTIVITY
1	CEO Americas	Initiates discourse at board level on approaches to expanding electronic trading business
2	Executive board	Decides to create separate corporate entity ETC to allow for the establishment of a new culture
3	Outside consultant	E-commerce discourse, ways in which to expand the domain of the business
4	Executive board	Discussion of corporate realignment of mission, goals, and objectives
5	CEO/senior management	Establishes strategic direction with senior management
6	Senior management/middle management	Meet to discuss and negotiate details of the procedures to implement
7	Middle management/operations communities	Meet with operations communities to discuss impact on day-to-day processes and procedures

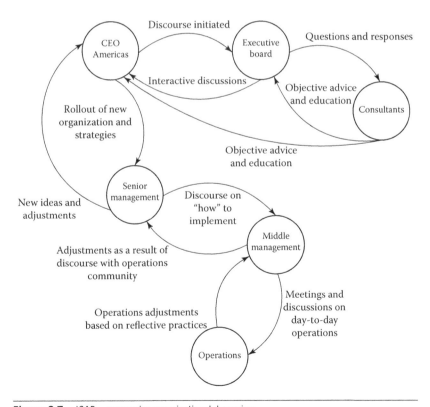

Figure 8.7   ICAP—responsive organizational dynamism.

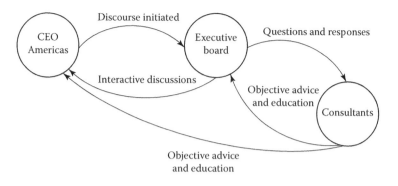

Figure 8.8 ICAP—community of practice.

on the very nomenclature applied to how technology was defined and procedures adopted among the international organizations within ICAP. That is why outside consultation could be effective as it provided independent and objective input that could foster the integration of culture-based concepts of technology, strategy, and ROI. Key to understanding the role of executive communities of practice is their overall importance to organizational learning growth. Very often we have heard, "Can we create productive discourse if the executive team cannot discuss and agree on issues themselves?" Effectively, ICAP created this community to ensure consistency among all the levels within the business. Consistent with the responsive organizational dynamism arc, learning in this community was at the "system" or organizational level, as opposed to being based on specific events like Y2K. These concerns had a broader context, and they affected both short- and long-term issues of business strategy and culture.

Another community of practice was the operations management team, which was the community responsible for transforming strategy into a realistic plan of strategic implementation. This team consisted of three levels (Figure 8.9). We see in this community of practice that the CEO was common to both this community and the executive community of practice. His participation in both provided the consistency and discourse that pointed to three valuable components:

1. The CEO could accurately communicate decisions reached at the board level to the operations management team.

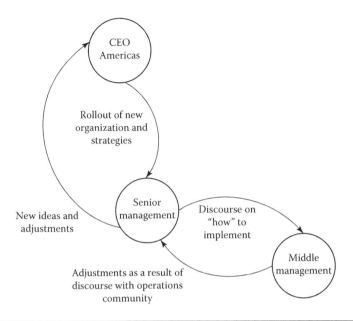

Figure 8.9   ICAP—community of practice interfaces.

2. The operations team could provide important input and suggestions to the CEO, who could then provide this information to the executive community.

3. The CEO interacted in different ways between the two communities of practice. This was critical because the way things were discussed, the language used, and the processes of consensus were different in each community.

The operations management community was not at the detailed level of implementation; rather, it was at the conceptual one. It needed to embrace the strategic and cultural outcomes discussed at the executive community, suggest modifications if applicable, and eventually reach consensus within the community and with the executive team. The operations management community, because of its conceptual perspectives, used more organizational learning methods as opposed to individual techniques. However, because of their relationship with operations personnel, they did participate in individual reflective practices. Notwithstanding their conceptual nature, event-driven issues were important for discussion. That is why middle management needed to be part of this community, for without their input, conceptual foundations for implementing change may very well have flaws.

Middle management participated to represent the concrete pieces and the realities for modifications to conceptual arguments. As such, middle managers could indirectly affect the executive board community since their input could require change in the operations management community, which in turn could foster the need for change requests back to the board. This process provides the very essence of why communities of practice need to work together, especially with the dynamic changes that can occur from technological innovations.

The third community of practice at ICAP was at the operations or implementation tier. It consisted of the community of staff that needed to transition conceptual plans into concrete realities. To ensure that conceptual ideas of implementation balanced with the concrete events that needed to occur operationally, middle managers needed to be part of both the operations management, and implementation communities, as shown in Figure 8.10.

Because of the transitory nature of this community, it was important that both organizational learning and individual learning occurred simultaneously. Thus, it was the responsibility of middle managers to provide the transition of organizational-based ideas to the event and concrete level so that individuals understood what it ultimately meant to the operations team. As one would expect, this level operated on individual attainment, yet through the creation of a community of practice, ICAP could get its operations members to begin to think more at the conceptual level. This provided management with the opportunity to discuss conceptual and system-level ideas with

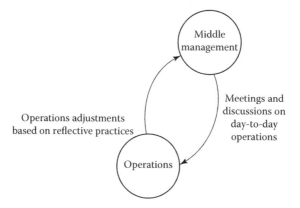

Figure 8.10   Middle-management community of practice at ICAP.

operations personnel. Operations personnel could review them and, under a managed and controlled process, could reach consensus. That is, changes required by the implementation community could be represented to the operations management community through middle management. If middle management could, through discourse and language, reach consensus with the operations management community, then the CEO could bring them forth to the executive community for further discussion. We can see this common thread concept among communities of practice as a logical process among tiers of operations and management and one that can foster learning maturation, as identified in the responsive organizational dynamism arc. This is graphically shown in Figure 8.11.

Figure 8.11 shows the relationships among the three communities of practice at ICAP and how they interacted, especially through upward feedback using common threads of communication. Thus, multiple communities needed to be linked via common individuals to maintain threads of communication necessary to support

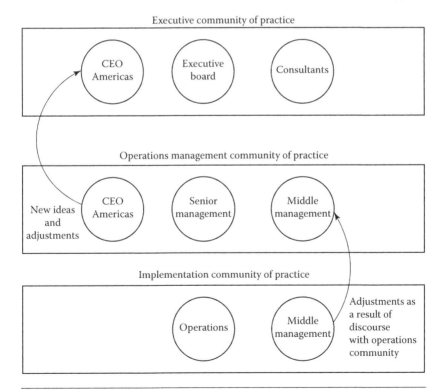

Figure 8.11   ICAP—COP common threads.

responsive organizational dynamism and learning across organizational boundaries.

Another important observation is the absence of independent consultants from the operations management and implementation communities of practice. This does not suggest that consultants were not needed or used by these communities. The independent consultant in the executive community provides organizational-level learning, as opposed to the consultant who is, for example, a specialist in database design or training.

This case study provides an example of how an international firm dealt with the effects of technology on its business. The CEO, Stephen McDermott in this case, played an important role, using many forms of responsive organizational dynamism, in managing the organization through a transformation. His experience fostered the realization that CEOs and their boards need to reinvent themselves on an ongoing basis. Most important, this case study identified the number of communities of practice that needed to participate in organizational transformation. The CEO continued to have an important role; in many ways, McDermott offered some interesting advice for other chief executives to consider:

1. The perfect time may or may not exist to deal with changes brought on by technology. The CEO may need to just "dive in" and serve as a catalyst for change.
2. Stay on course with the fundamentals of business and do not believe everything everyone tells you; make sure your business model is solid.
3. Trust that your abilities to deal with technology issues are no different from managing any other business issue.

As a result of the commitment and the process for adapting technology at ICAP, it has realized many benefits, such as the following:

- *Protection of tacit knowledge*: By incorporating the existing trading system, ICAP was able to retain the years of experience and expertise of its people. As a result, ICAP developed an electronic system that better served the needs of broker users; this ability gave it an advantage over competitor systems.

- *Integrated use*: The combination of the new system and its compatibility with other ICAP legacy systems enabled the organization to continue to service the core business while increasing access for new clients. This resulted in a reduction of costs and an increase in its user base.
- *Transformation of tacit knowledge to explicit product knowledge*: By providing an infrastructure of learning and strategic integration, ICAP was able to bridge a wide range of its employees' product knowledge, particularly of those outside IT with a specific understanding of trading system design, and to transform their tacit knowledge into explicit value that was used to build on to the existing trading systems.
- *Flexibility*: Because multiple communities of practice were formed, IT and non-IT cultures were able to assimilate. As a result, ICAP was able to reduce its overall development time and retain the functionality necessary for a hybrid voice and electronic trading system.
- *Expansion*: Because of the assimilation of cultures, ICAP was able to leverage its expertise so that the design of the electronic system allowed it to be used with other third-party trading systems. For example, it brought together another trading system from ICAP in Europe and enabled concurrent development in the United States and the United Kingdom.
- *Evolution*: By incorporating existing technology, ICAP continued to support the core business and gradually introduced new enhancements and features to serve all of its entities.
- *Knowledge creation*: By developing the system internally, ICAP was able to increase its tacit knowledge base and stay current with new trends in the industry.

ICAP went on to evolve its organization as a result of its adoption of technology and its implementation of responsive organizational dynamism. The company reinvented itself again. McDermott became the chief operating officer (COO) for three business units in the Americas; all specific business lines, yet linked by their integrated technologies and assimilated cultures. In addition, ICAP purchased a competitor electronic trading product and assimilated these combined technologies into a new organization. Business revenues rose at that time from $350 million to over $1 billion four years later. The company also had more than 2,800 staff members and operated from 21 offices worldwide. Much has been attributed to ICAP's investment in electronic trading systems and other emerging technologies.

Five Years Later

I returned to meet with Stephen McDermott almost five years after our original case study. Many of the predictions about how technology would affect the business had indeed become reality. In 2010, technology at ICAP had become the dominant component of the business. The C brokers had all but disappeared, with the organization now consisting of two distinct divisions: voice brokers and electronic brokers. The company continued to expand by acquiring other smaller competitors in the technology space. The electronic division now consisted of three distinct divisions from these acquisitions, with ETC just one of those divisions. In effect, the expansion led to more specialization and leveraging of technology to capture larger parts of various markets.

Perhaps the unseen reality was how quickly technology became a commodity. As McDermott said to me, "Everybody (our competitors) can do it; it's now all about your business strategy." While the importance of strategy was always part of McDermott's position, the transition from product value to market strategy was much more transformative on the organization's design and how it approached the market. For example, the additional regulatory controls on voice brokering actually forced many brokers to move to an electronic interface, which reduced liability between the buyer and the broker. McDermott also emphasized how "technology has created overnight businesses," forcing the organization to understand how technology could provide new competitive advantages that otherwise did not exist. Today, 50% of the trading dollars, some $2 trillion, occurs over electronic technology-driven platforms. Undoubtedly, these dynamic changes, brought on by technological dynamism, continue to challenge ICAP on how they strategically integrate new opportunities and how the organization must adapt culturally with changes in individual roles and responsibilities.

HTC

HTC (a pseudoacronym) is a company that provides creative business services and solutions. The case study involving HTC demonstrates that changes can occur when technology reports to the appropriate level in an organization. This case study offers the example of a company with a CEO who became an important catalyst in the successful vitalization of IT. HTC is a company of approximately 700 employees across 16 offices. The case involves studying the use of a new application that directly affected some 200 staff people.

The company was faced with the challenge of providing accurate billable time records to its clients. Initial client billings were based on project estimates, which then needed to be reconciled with actual work performed. This case turned out to be more complex than expected. Estimates typically represented the amount of work to which a client agreed. Underspending the budget agreed to by the client, however, could lead to lost revenue opportunities for the firm. For example, if a project was estimated at 20 hours, but the actual work took only 15, then most clients would seek an additional five hours of service because they had already budgeted the total 20 hours. If the reconciliation between hours budgeted and hours worked was significantly delayed, clients might lose their window of opportunity to spend the remaining five hours (in the example situation). Thus, the incapacity to provide timely reporting of this information resulted in the actual loss of revenue, as well as upset clients. If clients did not spend their allocated budget, they stood to lose the amount of the unused portion in their future budget allocations. Furthermore, clients had expectations that vendors were capable of providing accurate reporting, especially given that present-day technology could automate the recording and reporting of this information. Finally, in times of a tight economy, businesses tend to manage expenditures more closely and insist on more accurate record keeping than at other times.

The objective at HTC was to transform its services to better meet the evolving changes of its clients' business requirements. While the requirement for a more timely and accurate billing system seems straightforward, it became a greater challenge to actually implement than it otherwise seemed.

The first obstacle for HTC to overcome was the clash between this new requirement and the existing ethos, or culture of the business.

HTC provided creative services; 200 of its staff members were artistically oriented and were uncomfortable with focusing on time-based service tracking; they were typically engrossed in the creative performance required by their clients. Although it would seem a simple request to track time and enter it each day, this projected change in business norms became a significant barrier to its actual implementation. Project managers became concerned that reporting requirements would adversely affect performance, and thus, inevitably hurt the business. Efforts to use blunt force—do it or find another job — were not considered a good long-term solution. Instead, the company needed to seek a way to require the change while demonstrating the value of focusing on time management.

Many senior managers had thought of meeting with key users to help determine a workable solution, but they were cognizant of the fact that such interactive processes with the staff do not always lead to agreement on a dependable method of handling the problem. This is a common concern among managers and researchers working in organizational behavior. While organizational learning theorists advocate this mediating, interactive approach, it may not render the desired results in time and can even backfire if staff members are not genuinely willing to solve the problem or if they attempt to make it seem too difficult or a bad idea. The intervention of the CEO of HTC, together with the change in time reporting methods, directly involving IT, made a significant difference in overcoming the obstacle.

IT History at HTC

When I first interviewed the CEO, I found that she had little direct interaction with the activities of the IT department. IT reported to the CFO, as in many companies, because it was seen as an operational support department. However, the CEO subsequently became aware of certain shortfalls associated with IT and with its reporting structure. First, the IT department was not particularly liked by other departments. Second, the department seemed incapable of implementing software solutions that could directly help the business. Third, the CFO did not possess the creativity beyond accounting functions to provide the necessary leadership needed to steer the activities of IT in a more fruitful direction. As a result, the CEO decided that the IT department should report directly to her. She was also concerned that IT needed a more senior manager and hired a new chief technology officer (CTO).

Interactions of the CEO

My research involving 40 chief executives showed that many executives are unsure about what role they need to take with their chief IT managers. However, the CEO of HTC took on the responsibility to provide the financial support to get the project under way. First, the CEO made it clear that a solution was necessary and that appropriate funds would be furnished to get the project done. Second, the new CTO was empowered to assess the needs of the business and the staff, and to present a feasible solution for both business and cultural adaptation needs.

The CEO was determined to help transform the creative-artistic service business into one that would embrace the kinds of controls that were becoming increasingly necessary to support clients. Addressing the existing lag in collecting time records from employees, which directly affected billing revenue, seemed like the logical first step for engaging the IT department in the design and implementation of new operating procedures and cultural behavior.

Because middle managers were focused on providing services to their clients, they were less concerned with the collection of time sheets. This need was a low priority of the creative workers of the firm. Human resources (HR) had been involved in attempting to address the problem, but their efforts had failed. Much of this difficulty was attributed to an avoidance by middle managers of giving ultimatums as a solution; that is, simply demanding that workers comply. Instead, management subsequently became interested in a middle-ground approach that could possibly help departments realize the need to change and to help determine what the solution might be. The initial thinking of the CEO was to see if specialized technology could be built that would (1) provide efficiency to the process of recording time, and (2) create a form of controls that would require some level of compliance.

With the involvement of the CEO, the embattled IT department was given the authority to determine what technology could be employed to help the situation. The existing application that had been developed by the IT department did not provide the kind of ease of use and access that was needed by operations. Previous attempts to develop a new system, without the intervention of the CEO, had failed for a number of reasons. Management did not envision the potential solution that software was capable of delivering. It was not motivated in getting the requisite budget support; no one was in a position to champion it, to allocate the needed budget. Ultimately, management individuals were not convinced of the importance of providing a better solution.

The Process

The new CTO determined that there was a technological solution that could provide greater application flexibility, while maintaining its necessary integrity, through the use of the existing e-mail system. The application would require staff to enter their project time spent before signing on to the e-mail system. While this procedure might be seen as a punishment, it became the middle-ground solution for securing compliance without dramatically dictating policy. There was initial rejection of the procedure by some of the line managers, but it was with the assistance of the CEO, who provided the necessary support and enforcement, that the new procedure took hold. This enforcement became crucial when certain groups asked to be excluded from the process. The CEO made it clear that all departments were expected to comply.

The application was developed in three months and went into pilot implementation. The timely delivery of the application by the IT department gave IT its first successful program implementation and helped change the general view of IT among its company colleagues. It was the first occasion in which IT had a leadership role in guiding the company to a major behavioral transformation. Another positive outcome that resulted from the transition occurred in the way that resistance to change was managed by the CTO. Simply put, the creative staff was not open to a structured solution. The CTO's response was to implement a warning system instead of immediately disallowing e-mail access. This procedure was an important concession as it allowed staff and management to deal with the transition, to meet them halfway.

Transformation from the Transition

After the pilot period, the application was implemented firm-wide. The results of this new practice have created an interesting internal transformation: IT is now intimately engaged in working on new enhancements to the time-recording system. For instance, a "digital dashboard" is now used to measure performance against estimates. More important, however, are the results of the new application. The firm has shown substantial increases in revenue because its new time-recording system enabled it to discover numerous areas in which it was underbilling its clients. Its clients, on the other hand, are happier to receive billing statements that can demonstrate more accurately than before just how time was spent on their projects. Hence, the IT-implemented solution proved beneficial not only to the client but also to the firm.

Notwithstanding the ultimate value of utilizing appropriate technology and producing measurable outcomes, IT has also been able to assist in developing and establishing a new culture in the firm. Staff members are now more mindful and have a greater sense of corporate-norm responsibility than they did before. They have a clearer understanding of the impact that recording their time will have and of how this step ultimately contributes to the well-being of the business. Furthermore, the positive results of the new system have increased attention on IT spending. The CEO and other managers seek new ways in which technology can be made to help them; this mindset has been stressed further down to operating departments. The methods of IT evaluation have also evolved. There is now a greater clarification of technology benefits, a better articulation of technology problems, less trial and error, and more time spent on understanding how to use the technology better.

Another important result from this project has been the cascading effect of the financial impact. The increased profits have required greater infrastructure capacity. A new department was created with five new business managers whose responsibility it is to analyze and interpret the time reports so that line managers, in turn, can think of ways to generate greater profit through increased services. The project, in essence, has merged the creative performance of the firm with new business initiatives, resulting in a higher ROI.

In analyzing the HTC case study, we see many organizational learning techniques that were required to form a new community that could assimilate multiple cultures. However, while the organization saw the need, it could not create a process without an advocate. This champion was the CEO, who had the ability to make the salient organizational changes and act as a catalyst for the natural processes that HTC hoped to achieve. This case also provides direction on the importance of having the right resource to lead IT. At HTC, this person was the CTO; in actuality, this has little bearing on the overall role and responsibilities that were needed at HTC. At HTC, it became more apparent to the CEO that she had the wrong individual running the technology management of her firm. Only the CEO in this situation was able to foster the initial steps necessary to start what turned out to be a more democratic evolution of using technology in the business.

Companies that adapt to technological dynamism find that the existing leadership and infrastructure may need to be enhanced or replaced as well as reorganized, particularly in terms of reporting structure. This case supports the notion that strategic integration may indeed create the need for more cultural assimilation. One question to ask, is why the CEO waited so long to make the changes. This was not a situation of a new CTO who inherited resources. Indeed, the former CTO was part of her regime. We must remember that CEOs typically concentrate on driving revenue. They hope that what are considered "back-end" support issues will be handled by other senior managers. Furthermore, support structures are measured differently and from a specific frame of reference. I have found that CEOs intervene in supporter departments only when there are major complaints that threaten productivity, customer support, sales, and so on. The other threat is cost, so CEOs will seek to make supporter departments more efficient. These activities are consistent with my earlier findings regarding the measurement and role of supporter departments.

In the case of HTC, the CEO became more involved because of the customer service problems, which inevitably threatened revenues.

On her review of the situation, she recognized three major flaws in the operation:

- The CFO was not in a position to lead the organizational changes necessary to assimilate a creative-based department.
- Technology established a new strategy (strategic integration), which necessitated certain behavioral changes within the organization (cultural assimilation). The creative department was also key to make the organizational transition possible.
- The current CTO did not have the management and business skills that were necessary to facilitate the integration of IT with the rest of the organization.

HTC provides us with an interesting case of what we have defined as responsive organizational dynamism, and it bears some parallels to the Ravell study. First, like Ravell, the learning process was triggered by a major event. Second, the CTO did not dictate assimilation but rather provided facilitation and support. Unlike Ravell, the CEO of the organization was the critical driver to initiate the project. Because of the CEO's particular involvement, organizational learning started at the top and was thus system oriented. At the same time, the CTO understood that individual event-driven learning using reflective practices was critical to accomplish organizational transformation. In essence, the CTO was the intermediary between organizational-level and individual-level learning. Figure 8.12 depicts this relationship.

Five Years Later

HTC has been challenged because of the massive changes that advertising companies have faced over this timeframe, particularly with the difficulty of finding new advertising revenue sources for their clients. The CEO has remained active in technology matters, and there has also been turnover in the CTO role at the company. The CEO has been challenged to find the right fit—a person who can understand not only the technology but also the advertising business. With media companies taking over much of the advertising space, the CEO clearly recognizes the need to have a technology-driven market strategy. Most important is the dilemma of how to transform what

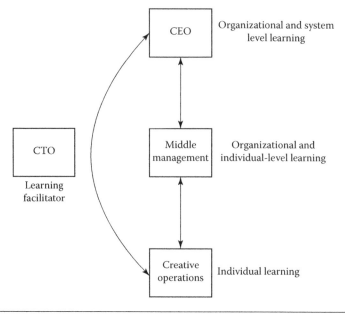

Figure 8.12 HTC—Role of the CTO as an intermediary.

was once a "paper" advertising business to what has become a lower-cost media market. "Advertising companies need to do more business just to keep the same revenue stream and that is a big challenge in today's volatile market," the CEO stated. The time-recording system has gone through other changes to provide what are known as *value-added services*, not necessarily tied to time effort, but rather, the value of the output itself.

The experience at HTC shows the importance of executive participation, not just sponsorship. Many technology projects have assumed the need for executive sponsorship. It is clear to me that this position is obsolete. If the CEO at HTC had not become involved in the problem five years ago, then the organization would not be in the position to embrace the newest technology dynamism affecting the industry. So, the lessons learned from this case, as well as from the Ravell case, are that all levels of the organization must be involved, and that executives must not be sacred. Responsive organizational dynamism, and the use of organizational learning methods to develop staff, remains key concepts for adapting to market changes and ensuring economic survival.

Summary

This chapter has provided three case studies that show the ways technology and organizational learning operate and lead to results through performance. The Siemens example provided us with an opportunity to see a technology executive formulate relationships, form multiple communities of practice, and create an infrastructure to support responsive organizational dynamism. This case provides a method in which IT can offer a means of handling technology as new information and, through the formation of communities of practice, it can generate new knowledge that leads to organizational transformation and performance.

The case study regarding ICAP again shows why technology, as an independent variable, provides an opportunity, if taken, for an international firm to move into a new competitive space and improve its competitive advantage. ICAP was only successful because it understood the need for organizational learning, communities of practice, and the important role of the CEO in facilitating change. We also saw why independent consultants and executive boards need to participate. ICAP symbolizes the ways in which technology can change organizational structures and cultural formations. Such changes are at the very heart of why we need to understand responsive organizational dynamism. The creation of a new firm, ETC, shows us the importance of these changes. Finally, it provides us with an example of how technology can come to the forefront of an organization and became the major driver of performance.

HTC, on the other hand, described two additional features of how responsive organizational dynamism can change internal processes that lead to direct returns. The CEO, as in the ICAP case, played an important, yet different, role. This case showed that the CTO could also be used to facilitate organizational learning, becoming the negotiator and coordinator between the CEO, IT department, and creative user departments.

All three of these cases reflect the importance of recognizing that most technology information exists outside the organization and needs to be integrated into existing cultures. This result is consistent with the findings of Probst et al. (1998), which show that long-term sustained competitive advantage must include the "incorporation and integration of information available outside the borders of the company" (p. 247). The reality is that technology, as an independent and outside variable, challenges organizations in their abilities to absorb external information, assimilate it into their cultures, and inevitably apply it to their commercial activities as a function of their existing knowledge base.

These case studies show that knowledge creation most often does not get created solely by individuals. It is by using communities of practice that knowledge makes its way into the very routines of the organization. Indeed, organizational learning must focus on the transformation of individual skills into organizational processes that generate measurable outcomes. Probst et al. (1998) also shows that the development of organizational knowledge is mediated via multiple levels. Walsh (1995) further supports Probst et al.'s findings that there are three structures of knowledge development in an organization. The first is at the individual level; interpretation is fostered through reflective practices that eventually lead to personal transformation and increased individual knowledge. The second structure is at the group level; individual knowledge of the group is combined into a consensus, leading to a shared belief system. The third structure resides at the organizational level; knowledge emanates from the shared beliefs and the consensus of the groups, which creates organizational knowledge. It is important to recognize, however, that organizational knowledge is not established or created by combining individual knowledge. This is a common error, particularly among organizational learning practitioners. Organizational knowledge must be accomplished through social discourse and common language interactions so that knowledge can be a consensus among the communities of practice.

Each of the case studies supported the formation of tiers of learning and knowledge. The individuals in these cases all created multiple layers that led to structures similar to those suggested by scholars. What makes these cases so valuable is that technology represented the external knowledge. Technological dynamism forced the multiple structures from individual-based learning to organizational-level learning, and the unique interactions among the communities in each example generated knowledge leading to measurable performance outcomes. Thus, as Probst and Büchel (1996, p. 245) conclude, "Organizational learning is an increase in organizational knowledge base, which leads to the enhancement of problem-solving potential of a company."

However, these case studies also provide important information about the process of the interactions. Many tiered structures tend to be viewed as a sequential process. I have presented theories suggesting that knowledge management is conditioned either from the top-down, middle-up-down, or bottom-up. It has been my position that none of these processes should be seen as set procedures or methodologies. In each of these cases, as well as in the Ravell case, the flow of knowledge occurs differently and, in some ways, uniquely to the culture and setting of the organization. This suggests that each organization must derive its own process, adhering more to the concept of learning, management, and outcomes, as opposed to a standard system of how and when they need to be applied. Table 8.2 summarizes the different approaches of organizational learning of the three case studies.

Such is the challenge of leaders who aspire to create the learning organization. Technology plays an important role because, in reality, it tests the very notions of organizational learning theories. It also creates many opportunities to measure organizational learning, and its impact on performance. Indeed, technology is the variable that provides the most opportunity to instill organizational learning, and knowledge management in a global community.

Table 8.2 Summary of Organizational Learning Approaches

SUBJECT	SIEMENS	ICAP/ETC	HTC
Knowledge management participation	CIO as middle-up-down	Top-down from CEO and bottom-up from operations	Top-down from CEO and middle-up-down from CTO
Community of practices	President's Council CFO CIO advisory board	Executive Board Operations Management Implementation	CEO/CTO CTO operations
Participating entities	Presidents CFOs Global CIO Corporate CIOs Regional CIOs Operating CIOs Central CIOs	Executive board Outside consultants CEO Senior management Middle management Operations	CEO CTO Middle management Creative operations
Common thread	Corporate CIO	CEO Senior management Middle management	CTO

The case studies also provided an understanding of the transformational process and the complexities of the relationships between the different learning levels. It is not a single entity that allows a company to be competitive but the combination of knowledge at each of the different tiers. The knowledge that exists throughout a company is typically composed of three components: processes, technology, and organization (Kanevsky & Housel, 1998). I find that, of these three components, technology is more variable than the others and, as stated many times in this book, at a dynamic and unpredictable fashion (that condition, called technological dynamism). Furthermore, the technology component has direct effects on the other two. What does this mean? Essentially, technology is at the core of organizational learning and knowledge creation.

This chapter has shown the different ways in which technology has been valued and how, through organizational learning, tacit knowledge is transformed into explicit knowledge, and used for competitive advantage. We have seen that not all of this value creation can be directly attributed to technology; in fact, this is rarely the case. Most value derived from technology is indirect, and it must be recognized by management as maximizing outcomes. Two of the case studies looked at the varying roles and responsibilities of the CEO. I believe their involvement was critical. Indeed, the conclusions reached from the Ravell case showed further support that the absence of the CEO will limit results. Furthermore, the CEO was crucial to sustaining organizational learning and the responsive organizational dynamism infrastructure.

Much has been written about the need to link learning to knowledge and knowledge to performance. This process can sometimes be referred to as a *value chain*. Kanevsky and Housel (1998) created what they call a "learning-knowledge-value spiral," comprised of six specific steps to creating value from learning and ultimately, changing product or process descriptions, as shown in Figure 8.13.

I have modified Figure 8.13 to include "technology"; that is, how technology affects learning, learning affects knowledge, and so on. Table 8.3 is a matrix that reflects the specific results, in each phase, for the three case studies.

Table 8.3 reflects the ultimate contribution that technology made to the learning-knowledge-value chain. I have also notated the ROI

Table 8.3 IT Contribution to the Learning-Knowledge-Value Chain

COMPANY	TECHNOLOGY	LEARNING	GENERATED KNOWLEDGE	PROCESS	PRODUCT	VALUE
Siemens	E-business	Communities of practice	Consensus across multiple communities on how to relate tacit knowledge about technology to strategic business processes.	90-Day "reinvention" life-cycle method.	Consolidated e-commerce Websites providing consistency of product and service offerings.	Leveraging of same clients; providing multiple product offerings to same client base. ROI: Indirect
ICAP	Electronic trading	CEO/executive committee Leveraging independent consultants. Group learning. Multiple communities of practice.	Ability to provide and integrate business and technology knowledge to create new product.	Establish new company, ETC, to support cultural assimilation and evolution.	Electronic trading.	Created most competitive product in the financial industry. Infiltration into new markets. ROI: Direct.
HTC	E-mail	CEO at organizational-level; individual learning using reflective practices.	Understanding how to integrate IT department with creative management group.	Establish new procedures for using e-mail to record client billable hours.	New client billing system.	Clients happy. More competitive. Additional revenues. ROI: Direct.

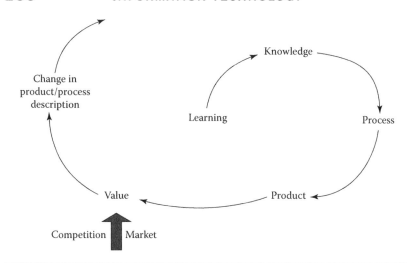

Figure 8.13   The learning-knowledge-value cycle. (From Kanevsky, V., et al. (Eds.), *Knowing in Firms: Understanding, Managing and Measuring Knowledge*, Sage, London, 1998, pp. 240–252.)

generated from each investment. It is interesting that two of the three cases generated identifiable direct revenue streams from their investment in technology.

This chapter has laid the foundation for Chapter 9, which focuses on the ways IT can maximize its relationship with the community and contribute to organizational learning. To accomplish this objective, IT must begin to establish best practices.

9

FORMING A CYBER SECURITY CULTURE

Introduction

Much has been written regarding the importance of how companies deal with cyber threats. While most organizations have focused on the technical ramifications of how to avoid being compromised, few have invested in how senior management needs to make security a priority. This chapter discusses the salient issues that executives must address and how to develop a strategy to deal with the various types of cyber attack that could devastate the reputation and revenues of any business or organization. The response to the cyber dilemma requires evolving institutional behavior patterns using organizational learning concepts.

History

From a historical perspective we have seen an interesting evolution of the types and acceleration of attacks on business entities. Prior to 1990, few organizations were concerned with information security except for the government, military, banks and credit card companies. In 1994, with the birth of the commercial Internet, a higher volume of attacks occurred and in 2001 the first nation-state sponsored attacks emerged. These attacks resulted, in 1997, in the development of commercial firewalls and malware. By 2013, however, the increase in attacks reached greater complexity with the Target credit card breach, Home Depot's compromise of its payment system, and JP Morgan's exposure that affected 76 million customers and seven million businesses. These events resulted in an escalation of fear, particularly in the areas of sabotage, theft of intellectual property, and stealing of money. Figure 9.1 shows the changing pace of cyber security

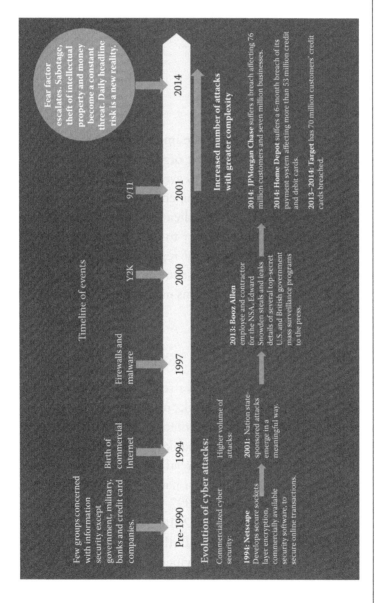

Figure 9.1 The changing pace of cyber security. (From Russell Reynolds Associates 2014 presentation.)

The conventional wisdom among cyber experts is that no business can be compromise proof from attacks. Thus, leaders need to realize that there must be (1) other ways beyond just developing new anti-software to ward off attacks, and (2) internal and external strategies to deal with an attack when it occurs. These challenges in cyber security management can be categorized into three fundamental components:

- Learning how to educate and present to the board of directors
- Creating new and evolving security cultures
- Understanding what it means organizationally to be compromised

Each of these components is summarized below

Talking to the Board

Board members need to understand the possible cyber attack exposures of the business. They certainly need regular communication from those executives responsible for protecting the organization. Seasoned security executives can articulate the positive processes that are in place, but without overstating too much confidence since there is always risk of being compromised. That is, while there may be exposures, C-level managers should not hit the panic button and scare the board. Typically, fear only instills a lack of confidence by the board in the organization's leadership. Most important is to always relate security to business objectives and, above all, avoid "tech" terms during meetings. Another important topic of discussion is how third-party vendors are being managed. Indeed, so many breaches have been caused by a lack of oversight of legacy applications that are controlled by third-party vendors. Finally, managers should always compare the state of security with that of the company's competitors.

Establishing a Security Culture

The predominant exposure to a cyber attack often comes from careless behaviors of the organization's employees. The first step to avoid poor employee cyber behaviors is to have regular communication with staff and establish a set of best practices that will clearly protect the business. However, mandating conformance is difficult and research has consistently supported that evolutionary culture change is best accomplished through relationship building, leadership by influence (as opposed to power-centralized management), and ultimately, a presence at most staff meetings. Individual leadership remains the most important variable when transforming the behaviors and practices of any organization.

Understanding What It Means to Be Compromised

Every organization should have a plan of what to do when security is breached. The first step in the plan is to develop a "risk" culture. What this simply means is that an organization cannot maximize protection of all parts of its systems equally. Therefore, some parts of a company's system might be more protected against cyber attacks than others. For example, organizations should maximize the protection of key company scientific and technical data first. Control of network access will likely vary depending on the type of exposure that might result from a breach. Another approach is to develop consistent best practices among all contractors and suppliers and to track the movement of these third parties (e.g., if they are merged/sold, disrupted in service, or even breached indirectly). Finally, technology executives should pay close attention to Cloud computing alternatives and develop ongoing reviews of possible threat exposures in these third-party service architectures.

Cyber Security Dynamism and Responsive Organizational Dynamism

The new events and interactions brought about by cyber security threats can be related to the symptoms of the dynamism that has been the basis of ROD discussed earlier in this book. Here, however, the digital world manifests itself in a similar dynamism that I will call *cyber dynamism*.

Managing cyber dynamism, therefore, is a way of managing the negative effects of a particular technology threat. As in ROD, *cyber* strategic integration and *cyber* cultural assimilation remain as distinct categories, that present themselves in response to cyber dynamism. Figure 9.2 shows the components of *cyber* ROD.

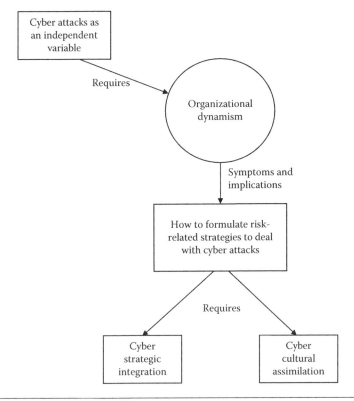

Figure 9.2   Cyber responsive organizational dynamism. (From Langer, A., *Information Technology and Organizational Learning: Managing Behavioral Change through Technology and Education*, CRC Press, Boca Raton, FL, 2011.)

Cyber Strategic Integration

Cyber strategic integration is a process that firms need to use to address the business impact of cyber attacks on its organizational processes. Complications posed by cyber dynamism, via the process of strategic integration, occurs when several new cyber attacks overlap and create a myriad of problems in various phases of an organization's ability to operate. Cyber attacks can also affect consumer confidence, which in turn hurts a business's ability to attract new orders. Furthermore, the problem can be compounded by reductions in productivity, which are complicated to track and to represent to management. Thus, it is important that organizations find ways to develop strategies to deal with cyber threats such as:

1. How to reduce occurrences by instituting aggressive organization structures that review existing exposures in systems.

2. What new threats exist, which may require ongoing research and collaborations with third-party strategic alliances?
3. What new processes might be needed to combat new cyber dynamisms based on new threat capabilities?
4. Creating systems architectures that can recover when a cyber breach occurs.

In order to realize these objectives, executives must be able to

- Create dynamic internal processes that can function on a daily basis, to deal with understanding the potential fit of new cyber attacks and their overall impact to the local department within the business, that is, to provide for change at the grass-roots level of the organization.
- Monitor cyber risk investments and determine modifications to the current life cycle of idea-to-reality.
- Address the weaknesses in the organization in terms of how to deal with new threats, should they occur, and how to better protect the key business operations.
- Provide a mechanism that both enables the organization to deal with accelerated change caused by cyber threats and that integrates them into a new cycle of processing and handling change.
- Establish an integrated approach that ties cyber risk accountability to other measurable outcomes integrating acceptable methods of the organization.

The combination of evolving cyber threats with accelerated and changing consumer demands has also created a business revolution that best defines the imperative of the strategic integration component of cyber ROD. Without action directed toward new strategic integration focused on cyber security, organizations will lose competitive advantage, which will ultimately affect profits. Most experts see the danger of breaches from cyber attacks as the mechanism that will ultimately require the integrated business processes to be realigned, thus providing value to consumers and modifying the customer-vendor relationship. The driving force behind this realignment emanates from cyber dynamisms, which serve as the principle accelerator of the change in transactions across all businesses.

Cyber Cultural Assimilation

Cyber cultural assimilation is a process that addresses the organizational aspects of how the security department is internally organized, its relationship with IT, and how it is integrated within the organization as a whole. As with technology dynamism, cyber dynamism is not limited only to cyber strategic issues, but *cultural* ones as well. A cyber culture is one that can respond to emerging cyber attacks, in an optimally informed way, and one that understands the impact on business performance and reputation.

The acceleration factors of cyber attacks require more dynamic activity within and among departments, which cannot be accomplished through discrete communications between groups. Instead, the need for diverse groups to engage in more integrated discourse and to share varying levels of cyber security knowledge, as well as business-end perspectives, requires new organizational structures that will give birth to a new and evolving business social culture.

In order to facilitate cyber cultural assimilation, organizations must have their staffs be more comfortable with a digital world that continues to be compromised by outside threats. The first question becomes one of finding the best structure to support a broad assimilation of knowledge about any given cyber threat. The second is about how that knowledge can best be utilized by the organization to develop both risk efforts and attack resilience. Business managers therefore need to consider cyber security and include the cyber staff in *all* decision-making processes. Specifically, cyber assimilation must become fundamental to the cultural evolution.

While many scholars and managers suggest the need to have a specific entity responsible for cyber security governance; one that is to be placed within the organization's operating structure, such an approach creates a fundamental problem. It does not allow staff and managers the opportunity to assimilate cyber security-driven change and understand how to design a culture that can operate under ROD. In other words, the issue of governance is misinterpreted as a problem of structural positioning or hierarchy when it is really one of cultural assimilation. As a result, many business solutions to cyber security issues often lean toward the prescriptive instead of the analytical in addressing the real problem.

Summary

This section has made the argument that organizations need to excel in providing both strategic and cultural initiatives to reduce exposure to cyber threats and ultimate security breaches. Executives must design their workforce to meet the accelerated threats brought on by cyber dynamisms. Organizations today need to adapt their staff to operate under the auspices of ROD by creating processes that can determine the strategic exposure of new emerging cyber threats and by establishing a culture that is more "defense ready." Most executives across industries recognize that cyber security has become one of the most powerful variables to maintaining and expanding company markets.

Organizational Learning and Application Development

Behavioral change, leading to a more resilient cyber culture, is just one of the challenges in maximizing protection in organizations. Another important factor is how to design more resilient applications that are better equipped to protect against threats; that is, a decision that needs to address exposure coupled with risk. The general consensus is that no system can be 100% protected and that this requires important decisions when analysts are designing applications and systems. Indeed, security access is not just limited to getting into the system, but applies to the individual application level as well. How then do analysts participate in the process of designing secure applications through good design? We know that many cyber security architectures are designed from the office of the chief information security officer (CISO), a new and emerging role in organizations. The CISO role, often independent of the chief information officer (CIO), became significant as a result of the early threats from the Internet, the 9/11 attacks and most recently the abundant number of system compromises experienced by companies such as JP Morgan Chase, SONY, Home Depot, and Target, to name just a few.

The challenge of cyber security reaches well beyond just architecture. It must address third-party vendor products that are part of the supply chain of automation used by firms, not to mention access to legacy applications that likely do not have the necessary securities built into the architecture of these older, less resilient technologies. This challenge has established the need for an enterprise cyber security solution that addresses the need of the entire organization. This approach would then target third-party vendor design and compliance. Thus, cyber security architecture requires integration with a firm's Software Development Life Cycle (SDLC), particularly within steps that include strategic design, engineering, and operations. The objective is to use a framework that works with all of these components.

Cyber Security Risk

When designing against cyber security attacks, as stated above, there is no 100% protection assurance. Thus, risks must be factored into the decision-making process. A number of security experts often ask business executives the question, "How much security do you want, and what are you willing to spend to achieve that security?"

Certainly, we see a much higher tolerance for increased cost given the recent significance of companies that have been compromised. This section provides guidance on how to determine appropriate security risks.

Security risk is typically discussed in the form of threats. Threats can be categorized as presented by Schoenfield (2015):

1. Threat agent: Where is the threat coming from, and who is making the attack?
2. Threat goals: What does the agent hope to gain?
3. Threat capability: What threat methodology, or type of approach is the agent possibly going to use?
4. Threat work factor: How much effort is the agent willing to put in to get into the system?
5. Threat risk tolerance: What legal chances is the agent willing to take to achieve his or her goals?

Table 9.1 is shown as a guideline.

Depending on the threat and its associated risks and work factors, it will provide important input to the security design, especially at the application design level. Such application securities in design typically include:

1. The user interface (sign in screen, access to specific parts of the application).

Table 9.1:   Threat Analysis

THREAT AGENT	GOALS	RISK TOLERANCE	WORK FACTOR	METHODS
Cyber criminals	Financial	Low	Low to medium	Known and proven

Source: Schoenfield, B.S.E., *Securing Systems: Applied Security Architecture and Threat Models,* CRC Press, Boca Raton, FL, 2015.

2. Command-line interface (interactivity) in online systems.
3. Inter-application communications. How data and password information are passed, and stored, among applications across systems.

Risk Responsibility

Schoenfield (2015) suggests that someone in the organization is assigned the role of the "risk owner." There may be many risk owners and, as a result, this role could have complex effects on the way systems are designed. For example, the top risk owner in most organizations today is associated with the CISO. However, many firms also employ a chief risk officer (CRO). This role's responsibilities vary.

But risk analysis at the application design level requires different governance. Application security risk needs involvement from the business and the consumer and needs to be integrated within the risk standards of the firm. Specifically, multiple levels of security often require users to reenter secure information. While this may maximize safety, it can negatively impact the user experience and the robustness of the system interface in general. Performance can obviously also be sacrificed, given the multiple layers of validation. There is no quick answer to this dilemma other than the reality that more security checkpoints will reduce user and consumer satisfaction unless cyber security algorithms become more invisible and sophisticated. However, even this approach would likely reduce protection. As with all analyst design challenges, the IT team, business users, and now the consumer must all be part of the decisions on how much security is required.

As my colleague at Columbia University, Steven Bellovin, states in his new book, *Thinking Security*, security is about a mindset. This mindset to me relates to how we establish security cultures that can enable the analyst to define organizational security as it relates to new and existing systems. If we get the analyst position to participate in setting security goals in our applications, some key questions according to Bellovin (2015) are:

1. What are the economics to protect systems?
2. What is the best protection you can get for the amount of money you want to spend?
3. Can you save more lives by spending that money?
4. What should you protect?
5. Can you estimate what it will take to protect your assets?
6. Should you protect the network or the host?
7. Is your Cloud secure enough?
8. Do you guess at the likelihood and cost of a penetration?
9. How do you evaluate your assets?
10. Are you thinking like the enemy?

The key to analysis and design in cyber security is recognizing that it is dynamic; the attackers are adaptive and somewhat unpredictable. This dynamism requires constant architectural change, accompanied with increased complexity of how systems become compromised. Thus, analysts must be involved at the conceptual model, which includes business definitions, business processes and enterprise standards. However, the analysts must also be engaged with the logical design, which comprises two sub-models:

1. *Logical architecture*: Depicts the relationships of different data domains and functionalities required to manage each type of information in the system.
2. *Component model*: Reflects each of the sub-models and applications that provide various functions in the system. The component model may also include third-part vendor products that interface with the system. The component model coincides, in many ways, with the process of decomposition.

In summary, the ROD interface with cyber security is more complex than many managers believe. Security is relative, not absolute, and thus leaders must be closely aligned with how internal cultures must evolve with changes environments.

Driver /Supporter Implications

Security has traditionally been viewed as a support function in most organizations, particularly when it is managed by IT staff. However, the recent developments in cyber threats suggest, as with other aspects of technology, that security too has a driver side.

To excel in the role of security driver, leaders must:

- Have capabilities, budgets and staffing levels, using benchmarks.
- Align even closer with users and business partners.
- Have close relationships with third parties.
- Extend responsibilities to include the growing challenges in the mobile workforce.
- Manage virtualized environments and third-party ecosystems.
- Find and/or develop cyber security talent and human capital.
- Have a strategy to integrate millennials with baby boomer and Gen X managers.

10

DIGITAL TRANSFORMATION AND CHANGES IN CONSUMER BEHAVIOR

Introduction

Digital transformation is one of the most significant activities of the early twenty-first century. Digital transformation is defined as "the changes associated with the applications of digital technology in all aspects of human society" (Stolterman & Fors, 2004, p. 689). From a business perspective, digital transformation enables organizations to implement new types of innovations and to rethink business processes that can take advantage of technology. From this perspective, digital transformation involves a type of reengineering, but one that is not limited to rethinking just how systems work together, but rather, that extends to the entire business itself. Some see digital transformation as the elimination of paper in organizations. Others see it as revamping a business to meet the demands of a digital economy. This chapter provides a link between digital transformation and what I call "digital reengineering." To explain this better, think of process reengineering as the generation that brought together systems in the way that they talked to one another—that is, the integration of legacy systems with new application that used more robust software applications.

The advent of digital transformation requires the entire organization to meet the digital demands of their consumers. For some companies, the consumer is another company (B2B, or business-to-business), that is, the consumer is a provider to another company that inevitably supports a consumer. For other businesses, their consumer is indeed the ultimate buyer. I will discuss the differences in these two types of consumer concepts later in this chapter. What is important from an IT perspective is that reengineering is no longer limited to just the needs of the internal user, but rather the needs of the businesses consumer as well. So, systems must change, as necessary, with the changes in consumer behavior. The challenge with doing this, of course, is that consumer needs are harder to obtain and understand, and can differ significantly among groups, depending on variables, such as ethnicity, age, and gender, to name just a few.

As a result, IT managers need to interact with the consumer more directly and in partnership with their business colleagues. The consumer represents a new type of user for IT staff. The consumer, in effect, is the buyer of the organization's products and services. The challenge becomes how to get IT more engaged with the buyer community, which could require IT to be engaged in multiple parts of the business that deals with the consumer. Below are six approaches, which are not mutually exclusive of each other:

1. *Sales/Marketing*: These individuals sell to the company's buyers. Thus, they have a good sense of what customers are looking for, what things they like about the business, and what they dislike. The power of the sales and marketing team is their ability to drive realistic requirements that directly impact revenue opportunities. The limitation of this resource is that it still relies on an internal perspective of the consumer; that is, how the sales and marketing staff perceive the consumer's needs.

2. *Third-party market analysis/reporting*: There are outside resources available that examine and report on market trends within various industry sectors. Such organizations typically have massive databases of information and, using various search and analysis tools, can provide a better understanding of the behavior patterns of an organization's consumers. These third parties can also provide reports that show how the organization stacks up against its competition and why consumers may be choosing alternative products. Unfortunately, if the data is inaccurate it likely will result in false generalizations about consumer behavior, so it is critical that IT digital leaders ensure proper review of the data integrity.

3. *Predictive analytics*: This is a hot topic in today's competitive landscape for businesses. Predictive analytics is the process of feeding off large data sets (*big data*) and predicting future

behavior patterns. Predictive analytics approaches are usually handled internally with assistance from third-party products or consulting services. The limitation is one of risk—the risk that the prediction does not occur as planned.

4. *Consumer support departments*: Internal teams and external vendors (outsourced managed service) have a good pulse on consumer preferences because they interact with them. More specifically, these department respond to questions, hande problems and get feedback from consumers on a regular basis. These support departments typically depend on applications to help the buyer. As a result, they are an excellent resource for providing up-to-date things that the system does not provide consumers. Unfortunately, consumer support organizations limit their needs to what they experience as opposed to what might be future trends of their consumers.

5. *Surveys*: IT and the business can design surveys (questionnaires) and send them to consumers for feedback. Using surveys can be of significant value in that the questions can target specific issues that the organization wants to address. Survey design and administration can be handled by third-party firms, which may have an advantage in that the questions are being forwarded from an independent source and one that does not identify the interested company. On the other hand, this might be considered a negative—it all depends on what the organization is seeking to obtain from the buyer.

6. *Focus groups*: This approach is similar to the use of a survey. Focus groups are commonly used to understand consumer behavior patterns and preferences. They are often conducted by outside firms. The differences between the focus group and a survey are (1) surveys are very quantitative based and use scoring mechanisms (Likert scales) to evaluate outcomes. Consumers sometimes may misinterpret the question thus resulting in distorted feedback, and (2) focus groups are more qualitative and allow IT digital leaders to engage with the consumer in two-way dialogues.

Figure 10.1 reflects a graphic depiction of the sources for understanding consumer behaviors and needs.

Table 10.1 further articulates the methods and deliverables that IT digital leaders should consider when developing system strategies.

Requirements without Users and without Input

Could it be possible to develop digital strategies and requirements for a system without user input or even consumer opinions? Could this be a reality for future design of strategic systems?

Perhaps we need to take a step back historically and think about trends that have changed the competitive landscape. Digital transformation may indeed be the most powerful agent of change in the history of business.

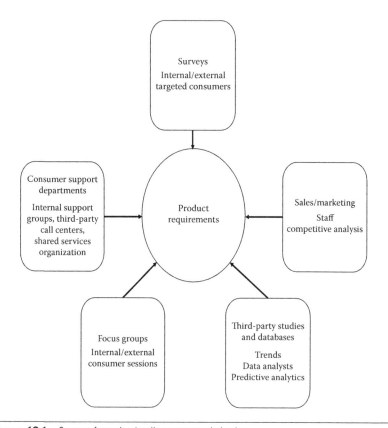

Figure 10.1   Sources for understanding consumer behavior.

Table 10.1 Langer's Methods and Deliverables for Assessing Consumer Needs

ANALYST'S SOURCES	METHODS	DELIVERABLES
Sales/ Marketing	Interviews	Should be conducted in a similar way to typical end user interviews. Work closely with senior sales staff. Set up interviews with key business stakeholders.
	Win/loss sales reviews	Review the results of sales efforts. Many firms hold formal win/loss review meetings that may convey important limitations of current applications and system capabilities.
Third-Party Databases	Document reports reviews	Obtain summaries of the trends in consumer behavior and pinpoint shortfalls that might exist in current applications and systems.
	Data analysis	Perform targeted analytics on databases to uncover trends not readily conveyed in available reports.
	Predictive analytics	Interrogate data by using analytic formulas that may enable predictive trends in consumer behavior.
Support Department	Interviews	Interview key support department personnel (internal and third party) to identify possible application deficiencies.
	Data/reports	Review call logs and recorded calls between consumers and support personnel to expose possible system deficiencies.
Surveys	Internal and external questionnaires	Work with internal departments to determine application issues when they support consumers. Use similar surveys with select populations of customers to validate and fine-tune internal survey results.
		Use similar surveys targeted to consumers who are not customers and compare results. Differences between existing customer base and non-customers may expose new trends in consumer needs.
Focus Groups	Hold internal and external sessions	Internal focus groups can be facilitated by marketing personnel. Select survey results, that had unexpected results or mixed feedback can be reviewed. Internal attendees should come from operations management and sales. External focus groups should be facilitated by a third-party vendor and held at independent sites. Discussions with customers should be compared with internal focus group results. Consumer focus groups should be facilitated by professional third-party firms.

We have seen large companies lose their edge. IBM's fall as the leading technology firm in the 1990s is an excellent example, when Microsoft overtook them. Yet Google was able to take the lead away from Microsoft, particularly in relation to analytical consumer computing. And what about the comeback Apple made with its new array of smart phone-related products? The question is, Why and how do these shifts in competitive advantage occur so quickly?

Technology continues to generate change and that change is typically referred to today as a "digital disruption." The challenge in disruption is the inability to predict what consumers want and need; furthermore, the consumer may not know! The challenge, then, is for IT digital leaders to forecast the changes that are brought about by technology disruptions. So, digital transformation is more about predicting consumer behavior and providing new products and services, which we hope consumers will want. This is a significant challenge for IT leaders, of course, given that the profession was built on the notion that good specifications accurately depicted what users want. Langer (1997) originally defined this as the "Concept of the Logical Equivalent." So, we may have created an oxymoron—how do we develop systems that the user cannot specify? Furthermore, requirements that depict consumer behavior are now further complicated by the globalization of business. Which consumer behavior are we attempting to satisfy and across what societal cultural norms? The reality is that new software applications will need to be built with some uncertainty. That is, some business rules may be vague and risks will need to be part of the process of system functionality. To see an example of designing systems based on uncertainty, we need only to analyze the evolution of the electronic spreadsheet. The first electronic spreadsheet, called VisiCalc, was introduced by a company called VisiCorp. It was designed for the Apple II and eventually the IBM personal computer. The electronic spreadsheet was not designed based on consumer input per se, rather on *perceived* needs by visionary designers who saw a need for a generic calculator and mathematical worksheet. VisiCorp took a risk by offering a product to the market that consumers would find useful. Of course, history shows that it was a very good risk. The electronic spreadsheet, which is now dominated by Microsoft's Excel product has gone through multiple product generations. The inventors of the electronic spreadsheet had a vision and the market responded favorably. Although VisiCorp's vision of the market need was correct, the first version was hardly 100% accurate of what consumers would want in a spreadsheet. For example, additional features, such as a database interface, three-dimensional spreadsheets to support budgeting and forward referencing, are all examples of responses from consumers that resulted in new product enhancements.

Allen and Morton (1994) established an excellent graphic depiction of the relationship between technology advancements and market needs (Figure 10.2)

Figure 10.2 shows an interesting life cycle of how product innovations relate to the creation of new products and services. The diagram reflects that innovations can occur as a result of new technology capabilities or inventions that establish new markets—like the electronic spreadsheet. On the other hand, the market can demand more features and functions the technology organizations or developers need to respond to that—like the upgrades made over the years to spreadsheet applications. Responding to market needs are what most organizations have practiced over the past 60 years, usually working with their end user populations (those internal users that supported the actual consumer). The digital revolution; however, is placing more emphasis on "generic" applications that resemble the object paradigm (one that requires applications to be able to fit into any business application). This trend will drive new and more advanced object-driven applications. These applications will reside in a more robust object functioning library that can dynamically link these modules together to form specific applications that can support mul consumer devices (what is now being called the "Internet of Things").

Another useful approach to dealing with consumer preferences is Porter's Five Forces Framework. Porter's framework consists of the following five components:

1. *Competitors*: What is the number of competitors in the market and what is the organization's position within the market?

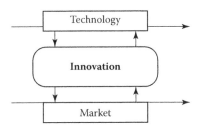

Figure 10.2 Technology, innovation, and market needs.

2. *New entrants*: What companies can come into the organization's space and provide competition?
3. *Substitutes*: What products or services can replace what you do?
4. *Buyers*: What alternatives do buyers have? How close and tight is the relationship between the buyer and seller?
5. *Suppliers*: What is the number of suppliers that are available, which can affect the relationship with the buyer and also determine price levels?

Porter's framework is graphically depicted in Figure 10.3.

Cadle et al. (2014) provide an approach to using Porter's model as part of the analysis and design process. Their approach is integrated with Langer's Analysis Consumer Methods in Table 10.2.

Concepts of the S-Curve and Digital Transformation Analysis and Design

Digital transformation will also be associated with the behavior of the S-curve. The S-curve has been a long-standing economic graph that depicts the life cycle of a product or service. The S-curve is shown in Figure 10.4

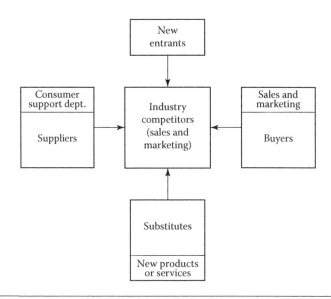

Figure 10.3 Porter's Five Forces Framework.

Table 10.2 Langer's Analysis Consumer Methods

PORTER'S FIVE FORCES	CADEL ET AL'S APPROACH	LANGER'S SOURCES OF INPUT
Industry competitors	How strong is your market share?	Third-party market studies
New entrants	New threats	Third-party market studies Surveys and focus groups
Suppliers	Price sensitivity and closeness of relationship.	Consumer support and end user departments
Buyers	Alternative choices and brand equity.	Sales/marketing team
Substitutes	Consumer alternatives	Surveys and focus groups Sales and marketing team Third-party studies

The left and lower portion of the S-curve represents a growing market opportunity that is likely volatile and exists where demand exceeds supply. As a result, the market opportunity is large and prices for the product are high. Thus, businesses should seek to capture as much of the market share at this time before competitors catch up. This requires the business to take more risk and assumes that the market will continue to demand the product. The shape of the S-curve suggests the life of this opportunity (the length of the x-axis represents the lifespan of the product).

As the market approaches the middle of the center of the S-curve, demand begins to equal supply. Prices start to drop and the market, in general, becomes less volatile and more predictable. The drop in price reflects the presence of more competitors. As a product or service approaches the top of the S, supply begins to exceed demand. Prices begin to fall and the market is said to have reached maturity. The uniqueness of the product or service is now approaching commodity.

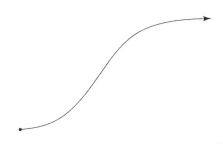

Figure 10.4 The S-curve.

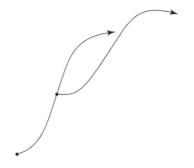

Figure 10.5 Extended S-curve.

Typically, suppliers will attempt to produce new features and functions to extend the life of the curve as shown in Figure 10.5

Establishing a new S-curve, then, extends the competitive life of the product or service. Once the top of the S-curve is reached, the product or service has reached the commodity level, where supply is much greater than demand. Here, the product or service has likely reached the end of its useful competitive life and should either be replaced with a new solution or considered for outsourcing to a third-party who can deliver the product at a very low price.

Langer's Driver/Supporter depicts the life cycle of any application or product as shown in Figure 10.6

Organizational Learning and the S-Curve

When designing a new application or system, the status of that product's S-curve should be carefully correlated to the source of the

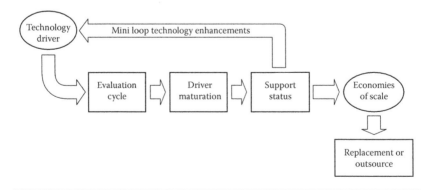

Figure 10.6 Langer's drive/supporter life cycle.

Table 10.3 S-Curve, Application Requirement Sources, and Risk

S-CURVE STATUS	ANALYSIS INPUT SOURCE	RISK FACTOR
Early S-curve	Consumer	High; market volatility and uncertainty.
High S-curve	Consumer	Lower; market is less uncertain as product becomes more mature.
	End users	Medium; business users have experience with consumers and can provide reasonable requirements.
Crest of the S-curve	End users	Low; business users have more experience as product becomes mature.
	Consumer	High; might consider new features and functions to keep product more competitive. Attempt to establish new S-curve.
End of S-curve	End user	None; seek to replace product or consider third-party product to replace what is now a legacy application. Also think of outsourcing application.

requirements. Table 10.3 reflects the corresponding market sources and associated risk factors relating to the dependability of requirements based on the state of the consumer's market. Leaders engaged in this process obviously need to have an abstract perspective to support a visionary and risk-oriented strategy. Table 10.3 includes the associated complexity of staff needed to deal with each period in the S-curve.

Communities of Practice

As stated in Chapter 4, Communities of Practice (COP) have been traditionally used as a method of bringing together people in organizations with similar talents, responsibilities and/or interests. Such communities can be effectively used to obtain valuable information about the way things work and what is required to run business operations. Getting such information strongly correlates to the challenges of obtaining dependable information from the consumer market. I discussed the use of surveys and focus groups earlier in this chapter, but COP is an alternative approach to bringing together similar types of consumers grouped by their interests and needs. In digital transformation we find yet another means of obtaining requirements by engaging in, and contributing to, the practices of specific consumer communities. This means that working with COP offers another way of developing relations with consumers to better understand their needs. Using this approach inside an organization, as we saw in Chapter 4, provides a means of better learning about issues by using a sustained method of remaining interconnected with specific business user groups, which can define what the organization really knows and contributes to the business that is typically not documented. IT digital leaders need to become engaged in learning if they are to truly understand what is needed to develop more effective and accurate software applications.

It seems logical that COP can provide the mechanism to assist IT digital leaders with an understanding of how business users and consumers behave and interact. Indeed, the analyst can target the behavior of the community and its need to consider what new organizational structures can better support emerging technologies. I have, in many ways, already established and presented what should be called the "community of IT digital leaders" and its need to understand how to restructure, in order to meet the needs of the digital economy. This new era does not lend itself to the traditional approaches to IT strategy, but rather to a more risk-based process that can deal with the realignment of business operations integrated with different consumer relationships.

The relationship, then, between COP and digital transformation is significant, given that future IT applications will heavily rely on informal inputs. While there may be attempts to computerize knowledge using predictive analytics software and big data, it will not be able to provide all of the risk-associated behaviors of users and consumers. That is, a "structured" approach to creating predictive behavior reporting, is typically difficult to establish and maintain. Ultimately, the dynamism from digital transformations creates too many uncertainties to be handled by sophisticated automated applications on how organizations will react to digital change variables. So, COP, along with these predictive analytics applications, provides a more thorough umbrella of how to deal with the ongoing and unpredictable interactions established by emerging digital technologies.

The IT Leader in the Digital Transformation Era

When we discuss the digital world and its multitude of effects on how business is conducted, one must ask how this impacts the profession of IT Leader. This section attempts to address the perceived evolution of the role.

1. The IT leader must become more innovative. While the business has the problem of keeping up with changes in their markets, IT needs to provide more solutions. Many of these solutions will not be absolute and likely will have short shelf lives. Risk is fundamental. As a result, IT leaders must truly become "business" leaders by exploring new ideas from the outside and continually considering how to implement the needs of the company's consumers. As a result, the business analyst will emerge as an idea broker (Robertson & Robertson, 2012) by constantly pursuing external ideas and transforming them into automated and competitive solutions. These ideas will have a failure rate, which means that companies will need to produce more applications than they will inevitably implement. This will certainly require organizations to spend more on software development.

2. Quality requirements will be even more complex. In order to keep in equilibrium with the S-curve the balance between quality and production will be a constant negotiation. Because applications will have shorter life cycles and there is pressure to provide competitive solutions, products will need to sense market needs and respond to them quicker. As a result, fixes and enhancements to applications will become more inherent in the development cycle after products go live in the market. Thus, the object paradigm will become even more fundamental to better software development because it provides more readily tested reusable applications and routines.

3. Dynamic interaction among users and business teams will require the creation of multiple layers of communities of practice. Organizations involved in this dynamic process must have autonomy and purpose (Narayan, 2015).

4. Application analysis, design, and development must be treated and managed as a living process; that is, it never ends until the product is obsolete (supporter end). So, products must continually develop to maturity.

5. Organizations should never outsource a driver technology until it reaches supporter status.

How Technology Disrupts Firms and Industries

The world economy is transforming rapidly from an analogue to a digital-based technology-driven society. This transformation requires businesses to move from a transactional relationship to one that that is "interactional" (Ernst & Young, 2012). However, this analogue to digital transformation, while essential for a business to survive in the twenty-first century, is difficult to accomplish. Langer's (2011) theory of responsive organizational dynamism (ROD), as discussed earlier in this book, is modified to show that successful adaptation of new digital technologies called *Digital Dynamisms* requires cultural assimilation of the people that comprise the organization.

Dynamism and Digital Disruption

The effects of digital dynamism can also be defined as a form of disruption or what is now being referred to as *digital disruption*. Specifically, the big question facing many enterprises is around how they can anticipate the unexpected threats brought on by technological advances that can devastate their business. There are typically two disruption factors:

1. A new approach to providing products and services to the consumer.
2. A strategy not previously feasible, now made possible using new technological capabilities.

Indeed, disruption occurs when a new approach meets the right conditions. Because technology shortens the time it takes to reach consumers, the changes are occurring at an accelerated and exponential pace. As an example, the table below shows the significant acceleration of the time it takes to reach 50 million consumers:

Radio	38 years
Television	13 years
Internet	4 years
Facebook	3.5 years
Twitter	9 months
Instagram	6 months
Pokémon GO	19 days

The speed of which we can accelerate change has an inverse effect on the length of time the effect lasts. We use the S-curve to show how digital disruption shortens the competitive life of new products and services. Figure 10.7 represents how the S-curve is shrinking along the x-axis, which measures the length or time period of the product/ service life.

Figure 10.7 essentially reflects that the life of a product or service is shrinking, thus enterprises have less time to capture a market opportunity and far less time to enjoy the length of its competitive success. As a result, business leaders are facing a world that is changing at an accelerating rate and trying to cope with understanding how new waves of "disruptive" technologies will affect their business. Ultimately, digital disruption shifts the way competitive forces deliver services, requires change in the way operations are managed and measured, and shortens the life of any given product or service success.

Critical Components of "Digital" Organization

A study conducted by Westerman et al. (2014), who interviewed 157 executives in fifty large companies, found four capabilities that were key to successful digital transformation:

1. *A unified digital platform*: Integration of the organization's data and processes across its department silos is critical. One reason why web-based companies gain advantage over traditional competitors is their ability to use analytics and customer personalization from central and integrated sources. Thus, the first step toward a successful digital transformation is for companies to invest in establishing central repositories of data and common applications that can access the information.

Figure 10.7 The shrinking S-curve.

This centralization of digital data is key to competing globally since firms must be able to move data to multiple locations and use that data in different contexts.

2. *Solution delivery*: Many traditional IT departments are not geared to integrating new processes into their legacy operations. A number of firms have addressed this problem by establishing independent "innovation centers" designed to initiate new digital ideas that are more customer solution oriented. These centers typically focus on how new mobile and social media technologies can be launched without disturbing the core technology systems that support the enterprise. Some of these initiatives include partnerships with high-tech vendors; however, a number of executives have shown concern that such alliances might result in dependencies because of the lack of knowledge inside the organization.

3. *Analytics capabilities*: Companies need to ensure that their data can be used for predictive analytics purposes. Predictive analytics provide actors with a better understanding of their consumer's behaviors and allow them to formulate competitive strategies over their competitors. Companies that integrate data better from their transactional systems can make more "informed and better decisions" and formulate strategies to take advantage of customer preferences and thus, turn them into business opportunities. An example is an insurance company initiative that concentrates on products that meet customer trends determined by examining their historical transactions across various divisions of the business. Analytics also helps organizations to develop risk models that can assist them to formulate accurate portfolios.

4. *Business and IT integration*: While the integration of the IT department with the business has been discussed for decades, few companies have achieved a desired outcome (Langer, 2016). The need for digital transformation has now made this integration essential for success and to avoid becoming a victim of disruption. True IT and business integration means more than just combining processes and decision making; but rather, the actual movement of personnel into business units so they can be culturally assimilated (Langer & Yorks, 2013).

Assimilating Digital Technology Operationally and Culturally

When considering how to design an organization structure that can implement digital technologies, firms must concentrate on how to culturally assimilate a new architecture. The importance of the architecture first affects the strategic integration component of ROD. Indeed, the actor-oriented architecture must be designed to be agile enough to react to increased changes in market demands. The consumerization of technology, defined as changes in technology brought on by increased consumer knowledge of how digital assets can reduce costs and increase competitive advantage, have created a continual reduction in the length of any new competitive products or services life. Thus, consumerization has increased what Eisenhardt and Bourgeouse (1988) define as "high-velocity" market conditions.

This dilemma drives the challenge of how organizations will cope to avoid the negative effects of digital disruption. There are four overall components that appear to be critical factors of autonomy from disruption:

1. Companies must recognize that speed and comfort of service can be more important than just the cost: our experience is that enterprises who offer multiple choices that allow consumers to choose from varying levels of service options are more competitive. The more personal the service option, the higher the cost. Examples can be seen in the airline industry where passengers have options for better seats at a higher price, or a new option being offered by entertainment parks that now provide less wait time on shorter lines, for higher paying customers. These two examples match the price with a desired service and firms that do not offer creative pricing options are prime for disruption.

2. Empower your workforce to try new ideas without over controls. Companies are finding that many young employees have new service ideas but are blocked from trying them because of the "old guard" in their management reporting lines. Line managers need to be educated on how to allow their staffs to quickly enact new processes, even though some of them may not be effective.

3. Allow employees and customer to have choice of devices. Traditionally IT departments desire to create environments where employees adhere to standard hardware and software structures. Indeed, standard structures make it easier for IT to support internal users and provide better security across systems. However, as technology has evolved, the relation between hardware and software, especially in mobile devices, has become more specialized. For example, Apple smartphones have proprietary hardware architectures that in many cases require different versions of application software as well as different security considerations than its major competitor, Samsung. With the consumerization of technology, these IT departments must now support multiple devices because both their customers and employees are free to select them. Therefore, it is important to allow staff to freely integrate company applications with their personal device choices.

4. Similar to (3), organizations who force staff to adhere to strict processes and support structures are exposed to digital disruption. Organizational structures that rely on technological innovation must be able to integrate new digital opportunities seamlessly into their current production and support processes. Specifically, this means having the ability to be agile enough to provide services using different digital capabilities and from different geographical locations.

Conclusion

This chapter has provided a number of different and complex aspects of digital transformation, its effects on how organizations are structured and how they need to compete to survive in the future. The technology executive is, by default, the key person to lead these digital transformation initiatives because of the technical requirements that are at the center of successfully completing these projects. As such, these executives must also focus on their own transformation as leaders that allows them to help form the strategic goals to meet the dynamic changes in consumer behavior.

11

INTEGRATING GENERATION Y EMPLOYEES TO ACCELERATE COMPETITIVE ADVANTAGE

Introduction

This chapter focuses on Gen Y employees who are also known as "digital natives" and "millennials." Gen Y employees possess the attributes to assist companies in transforming their workforce to meet the accelerated change in the competitive landscape. Most executives across industries recognize that digital technologies are the most powerful variable to maintaining and expanding company markets. Gen Y employees provide a natural fit for dealing with emerging digital technologies. However, success with integrating Gen Y employees is contingent upon baby boomer and Gen X management to adapt new leadership philosophies and procedures suited to meet the expectations and needs of these new workers. Ignoring the unique needs of Gen Y employees will likely result in an incongruent organization that suffers high turnover of young employees who will seek more entrepreneurial environments.

I established in Chapter 10 that digital transformation is at the core of change and competitive survival in the twenty-first century. Chapter 10 did not address the changes in personnel that are quickly becoming major issues at today's global firms. While I offered changes to organizational structures, I did not address the mixture of different generations that are at the fabric of any typical organization. This chapter is designed to discuss how these multiple generations need to "learn" how to work together to form productive and effective organizations that can compete in the digital economy. Furthermore, this chapter will address how access to human capital will change in the future and the different types of relationships that individuals will have with employers. For example, the "gig" economy will use non-traditional outside workers who will provide sources of talent for shorter-term employment needs. Indeed, the gig economy will require HR and IT leaders to form new and intricate employee relationships.

As discussed in Chapter 10, companies need to transform their business from analogue to one that uses digital technologies. Such transformation requires moving from a transactional relationship with customers to one that is more "interactional" (Ernst & Young, 2012). Completing an analogue to digital transformation, while essential for a business to survive in the twenty-first century, is difficult to accomplish. Responsive organizational dynamism (ROD) showed us that successful adaptation of new digital technologies requires strategic integration and cultural assimilation of the people that comprise the organization. As stated earlier, these components of ROD can be categorized as the essential roles and responsibilities of the organization that are necessary to utilize new technological inventions that can strategically be integrated within a business entity. The purpose here is to explore why Gen Y employees need to be integrated with baby boomers and Gen X staff to effectively enhance the success of digital transformation initiatives.

The Employment Challenge in the Digital Era

Capgemini and MIT (2013) research shows that organizations need new operating models to meet the demands of a digital-driven era. Digital tools have provided leaders with ways to connect at an unprecedented scale. Digital technology has allowed companies to invade other spaces previously protected by a business's "asset specificities" (Tushman & Anderson, 1997), which are defined as advantages enjoyed by companies because of their location, product access, and delivery capabilities. Digital technologies allow those specificities to be neutralized and thus, change the previous competitive balances among market players. Furthermore, digital technology accelerates this process, meaning that changes in market share occur very quickly. The research offers five key indicators that support successful digital transformation in a firm:

1. A company's strategic vision is only as effective as the people behind it. Thus, winning the minds of all levels of the organization is required.
2. To become digital is to be digital. Companies must have a "one-team culture" and raise their employees' digital IQ.
3. A company must address the scarcity of talented resources and look more to using Gen Y individuals because they have a more natural adaptation to take on the challenges of digital transformation.
4. Resistant managers are impediments to progress and can actually stop digital transformation.
5. Digital leadership starts at the top.

As stated in Chapter 10, Eisenhardt and Bourgeouis (1988) first defined dynamic changing markets as being "high-velocity." Their research shows that high-velocity conditions existed in the technology industry during the early 1980s in Silicon Valley, in the United States. They found that competitive advantage was highly dependent on the quality of people that worked at those firms. Specifically, they concluded that workers who were capable of dealing with change and less subjected to a centralized totalitarian management structure outperformed those that had more traditional hierarchical organizational structures. While "high-velocity" during the 1980s was unusual, digital disruption in the twenty-first century has made it a market norm.

The combination of evolving digital business drivers with accelerated and changing customer demands has created a business revolution that best defines the imperative of the strategic integration component of ROD. The changing and accelerated way businesses deal with their customers and vendors requires a new strategic integration to become a reality, rather than remain a concept without action. Most experts see digital technology as the mechanism that will require business realignment to create new customer experiences. The driving force behind this realignment emanates from digital technologies, which serve as the principle accelerator of the change in transactions across all business units. The general need to optimize human resources forces organizations to rethink and to realign business processes, in order to gain access to new business markets, which are weakening the existing "asset specificities" of the once dominant market leaders.

Gen Y Population Attributes

Gen Y or digital natives are those people who are accustomed to the attributes of living in a digital world and are 18–35 years old. Gen Y employees are more comfortable with accelerated life changes, particularly change brought on by new technologies. Such individuals, according to a number of commercial and academic research studies (Johnson Controls, 2010; Capgemini, 2013; Cisco, 2012; Saxena & Jain, 2012), have attributes and expectations in the workplace that support environments that are flexible, offer mobility, and provide collaborative and unconventional relationships. Specifically, millennial workers

- want access to dedicated team spaces where they can have emotional engagements in a socialized atmosphere;
- require their own space; that is, are not supportive of a "hoteling" existence where they do not have a permanent office or workspace;
- need a flexible life/work balance;
- prefer a workplace that supports formal and informal collaborative engagement.

Research has further confirmed that 79% of Gen Y workers prefer mobile jobs, 40% want to drive to work, and female millennials need more flexibility at work than their male counterparts. As a result of this data, businesses will need to compete to recruit and develop skilled Gen Y workers who now represent 25% of the workforce. In India, while Gen Y represents more than 50% of the working population, the required talent needed by businesses is extremely scarce.

Advantages of Employing Millennials to Support Digital Transformation

As stated, Gen Y adults appear to have many identities and capabilities that fit well in a digital-driven business world. Indeed, Gen Y people are consumers, colleagues, employees, managers, and innovators (Johnson Controls, 2010). They possess attributes that align with the requirements to be an entrepreneur, a person with technology savvy and creativity, someone who works well in a mobile environment, and is non-conformant enough to drive change in an organization. Thus, the presence of Gen Y personnel can help organizations to restrat-egize their competitive position and to retain key talent (Saxena & Jain, 2012). Furthermore, Gen Y brings a more impressive array of academic credentials than their predecessors.

Most important is Gen Y's ability to deal better with market change—which inevitably affects organizational change. That is, the digital world market will constantly require changes in organizational structure to accommodate its consumer needs. A major reason for Gen Y's willingness to change is its natural alignment with a company's customers. Swadzba (2010) posits that we are approaching the end of what he called the "work era" and moving into a new age based on consumption. Millennials are more apt to see the value of their jobs from their own consumption needs. Thus, they see employment as an act of consumption (Jonas & Kortenius, 2014). Gen Y employees therefore allow employers to acquire the necessary talent that can lead to better consumer reputation, reduced turnover of resources and, ulti-mately, increased customer satisfaction (Bakanauskiené et al., 2011). Yet another advantage of Gen Y employees is their ability to transform organizations that operate on a departmental basis into one that is based more on function; an essential requirement in a digital economy.

Integration of Gen Y with Baby Boomers and Gen X

The prediction is that 76 million baby boomers (born 1946–1964) and Gen X workers (born 1965–1984) will be retiring over the next 15 years. The question for many corporate talent executives is how to manage the transition in a major multigenerational workforce. Baby boomers alone still inhabit the most powerful leadership positions in the world. Currently, the average age of CEOs is 56, and 65% of all corporate leaders are baby boomers. Essentially, corporations need to produce career paths that will be attractive to millennials. Thus, the older generation needs to

- Acknowledge some of their preconceived perceptions of cur-rent work ethics that are simply not relevant in today's com-plex environments.
- Allow Gen Y to escalate in ranks to satisfy their ambitions and sense of entitlement.

- Implement more flexible work schedules, offer telecommuting, and develop a stronger focus on social responsibility.
- Support more advanced uses of technology, especially those used by Gen Yers in their personal lives.
- Employ more mentors to help Gen Y employees to better understand the reasons for existing constraints in the organizations where they work.
- Provide more complex employee orientations, more timely personnel reviews, and in general more frequent feedback needed by Gen Y individuals.
- Establish programs that improve the verbal communications skills of Gen Y workers that are typically more comfortable with nonverbal text-based methods of communication.
- Implement more continual learning and rotational programs that support a vertical growth path for younger employees.

In summary, it is up to the baby boomer and Gen X leaders to modify their styles of management to fit the needs of their younger Gen Y employees. The challenge to accomplish this objective is complicated, given the wide variances on how these three generations think, plan, take risks, and most important, learn.

Designing the Digital Enterprise

Zogby completed an interactive poll of 4,811 people on perceptions of different generations. 42% of the respondents stated that baby boomers would be remembered for their focus on consumerism and self-indulgence. Gen Y, on the other hand, are considered more self-interested, entitled narcissists who want to spend all their time posting "selfies" to Facebook. However, other facts offer an expanded perception of these two generations, as shown in Table 11.1

Research completed by Ernst and Young (2013) offers additional comparisons among the three generations as follows:

1. Gen Y individuals are moving into management positions faster due to retirements, lack of corporate succession planning, and their natural ability to use technology at work. Table 11.2 shows percentage comparisons between 2008 and 2013.

Table 11.1   Baby Boomers versus Gen Y

BABY BOOMERS	GEN Y
Married later and less children	Not as aligned to political parties
Spend lavishly	More civically engaged
More active and selfless	Socially active
Fought against social injustice, supported civil rights, and defied the Vietnam War	Cheerfully optimistic
Had more higher education access	More concerned with quality of life than material gain

The acceleration of growth to management positions among Gen Y individuals can be further illuminated in Table 11.3 by comparing the prior five-year period from 2003 to 2007.

2. While responders of the survey felt Gen X were better equipped to manage than Gen Y, the number of Gen Y managers is expected to double by 2020 due to continued retirements. Another interesting result of the research relates to Gen Y expectations from their employers when they become managers. Specifically Gen Y managers expect (1) an opportunity to have a mentor, (2) to receive sponsorship, (3) to have more career-related experiences, and (4) to receive training to build their professional skills.

3. Seventy-five percent of respondents that identified themselves as managers agree that managing the multiple generations is a significant challenge. This was attributed to different work expectations and the lack of comfort with younger employees managing older employees.

Table 11.4 provides additional differences among the three generations:

Table 11.2   Management Roles 2008–2013

Baby boomer (ages 49–67)	19%
Gen X (ages 33–48)	38%
Gen Y (18–32)	87%

Table 11.3   Management Roles 2003–2007

Baby boomer (ages 49–67)	23%
Gen X (ages 33–48)	30%
Gen Y (18–32)	12%

Table 11.4   Baby Boomer, Gen X and Gen Y Compared

BABY BOOMERS	GEN X	GEN Y
Seek employment in large established companies that provide dependable employment.	Established companies no longer a guarantee for lifetime employment. Many jobs begin to go offshore.	Seek multiple experiences with heavy emphasis on social good and global experiences. Re-evaluation of offshoring strategies.
Process of promotion is well defined, hierarchical and structured, eventually leading to promotion and higher earnings—concept of waiting your turn.	Process of promotion still hierarchical, but based more on skills and individual accomplishments. Master's degree now preferred for many promotions.	Less patience with hierarchical promotion policies. More reliance on predictive analytics as the basis for decision making.
Undergraduate degree preferred but not mandatory.	Undergraduate degree required for most professional job opportunities.	More focus on specific skills. Multiple strategies developed on how to meet shortages of talent. Higher education is expensive and concerns increase about the value of graduate knowledge and abilities.
Plan career preferably with one company and retire. Acceptance of a gradual process of growth that was slow to change. Successful employees assimilated into existing organizational structures by following the rules.	Employees begin to change jobs more often, given growth in the technology industry, and opportunities to increase compensation and accelerate promotion by switching jobs.	Emergence of a "gig" economy, and the rise of multiple employment relationships
Entrepreneurism was seen as an external option for those individuals desiring wealth and independence and willing to take risks.	Corporate executives' compensation dramatically increases, no longer requiring starting businesses as the basis for wealth.	Entrepreneurism promoted in Higher Education as the basis for economic growth, given the loss of jobs in the U.S.

Assimilating Gen Y Talent from Underserved and Socially Excluded Populations

The outsourcing of jobs outside of local communities to countries with lower employment costs has continued to grow during the early part of the twenty-first century. This phenomenon has led to significant social and economic problems, especially in the United States and in Western Europe as jobs continue to migrate to foreign countries where there are lower labor costs and education systems that provide more of the skills needed by corporations. Most impacted by the loss of jobs have been the underserved or socially excluded Gen Y youth populations. Indeed, the European average for young adult unemployment (aged 15–25) in 2013 was nearly 25%, almost twice the rate for their adult counterparts (Dolado, 2015). Much of the loss of local jobs can be attributed to expansion of the globalized economy, which has been accelerated by continued technological advancements (Wabike, 2014). Thus, the effects of technology gains have negatively impacted efforts toward social inclusion and social equality.

Langer, in 2003, established an organization called Workforce Opportunity Services (WOS), as a means of utilizing a form of action research using adult development theory to solve employment problems caused by outsourcing. Langer's approach is based on the belief that socially excluded youth can be trained and prepared for jobs in areas such as information technology that would typically be outsourced to lower labor markets. WOS has developed a talent-finding model that has successfully placed over 1400 young individuals in such jobs. Results of over 12 years of operation and research have shown that talented youth in disadvantaged communities do exist and that such talent can economically and socially contribute to companies (Langer, 2013). The following section describes the Langer Workforce Maturity Arc (LWMA), presents data on its effectiveness as a transformative learning instrument, and discusses how the model can be used as an effective way of recruiting Gen Y talent from underserved and socially excluded populations.

Langer Workforce Maturity Arc

The Langer Workforce Maturity Arc (LWMA) was developed to help evaluate socially excluded youth preparation to succeed in the workplace. The LWMA, initially known as the Inner-City Workplace Literacy Arc:

> charts the progression of underserved or 'excluded' individuals along defined stages of development in workplace culture and skills in relation to multiple dimensions of workplace literacy such as cognitive growth and self-reflection. When one is mapped in relation to the other (workplace culture in relation to stages of literacy assimilation), an Arc is created. LWMA traces the assimilation of workplace norms, a form of individual development. (Langer, 2003: 18)

The LWMA addresses one of the major challenges confronting an organization's HR group: to find talent from diverse local populations that can successfully respond to evolving business norms, especially those related to electronic and digital technologies. The LWMA provides a method for measuring the assimilation of workplace cultural norms and thus, can be used to meet the mounting demands of an increasingly global, dynamic, and multicultural workplace. Furthermore, if organizations are to attain acceptable quality of work from diverse employees, assimilation of socially or economically excluded populations must be evaluated based on (1) if and how individuals adopt workplace cultural norms, and (2) how they become integrated into the business (Langer, 2003). Understanding the relationship between workplace assimilation and its development can provide important information on how to secure the work ethic, dignity, solidarity, culture, cognition, and self-esteem of individuals from disadvantaged communities, and their salient contributions to the digital age.

Theoretical Constructs of the LWMA

The LWMA encompasses *sectors of workplace literacy* and *stages of literacy development*, and the arc charts business acculturation requirements as they pertain to disadvantaged young adult learners. The relationship between workplace assimilation and literacy is a challenging subject. A specific form of literacy can be defined as a social practice that requires specific skills and knowledge (Rassool, 1999). In this instance, workplace literacy addresses the effects of workplace practices and culture on the social experiences of people in their workday, as well as their everyday lives. We need to better understand how individual literacy in the workplace, which subordinates individuality to the demands of an organization, is formulated for diverse groups (Newman, 1999). Most important, are the ways in which one learns how to behave effectively in the workplace—the knowledge, skill, and attitude sets required by business generally, as well as by a specific organization. This is particularly important in disadvantaged communities, which are marginalized from the experiences of more affluent communities in terms of access to high-quality education, information technologies, job opportunities, and workplace socialization. For example, Friedman et al. (2014) postulate that the active involvement of parents in the lives of their children greatly impacts a student's chances of success. It is the absence of this activism that contributes to a system of social exclusion of youth. Prior to determining what directions to pursue in educational pedagogies and infrastructures, it is necessary to understand what workplace literacy requirements are present and how they can be developed for disadvantaged youth in the absence of the active support from families and friends.

The LWMA assesses individual development in six distinct *sectors* of workplace literacy:

1. *Cognition*: Knowledge and skills required to learn and complete job duties in the business world, including computational skills; ability to read, comprehend, and retain written information quickly; remembering and executing oral instructions; and critically examining data.

2. *Technology*: An aptitude for operating various electronic and digital technologies.

3. *Business culture*: Knowledge and practice of proper etiquette in the workplace including dress codes, telephone and in-person interactions, punctuality, completing work and meeting deadlines, conflict resolution, deference and other protocols associated with supervisors and hierarchies.

4. *Socio-economic values*: Ability to articulate and act upon mainstream business values, which shape the work ethic. Such values include independent initiative, dedication, integrity, and personal identification with career goals. Values are associated with a person's appreciation for intellectual life, cultural sensitivity to others, and sensitivity for how others view their role in the workplace. Individuals understand that they should make decisions based on principles and evidence rather than personal interests.

5. *Community and ethnic solidarity*: Commitment to the education and professional advancement of persons in ethnic minority groups and underserved communities. Individuals can use their ethnicity to explore the liberating capacities offered in the workplace without sacrificing their identity (i.e., they can assimilate workplace norms without abandoning cultural, ethnic, or self-defining principles and beliefs).

6. *Self-esteem*: The view that personal and professional success work in tandem, and the belief in one's capacity to succeed in both arenas. This includes a devotion to learning and self-improvement. Individuals with high self-esteem are reflective about themselves and their potential in business. They accept the realities of the business world in which they work and can comfortably confirm their business disposition, independently of others' valuations.

Each stage in the course of an individual's workplace development reflects an underlying principle that guides the process of adopting workplace norms and behavior. The LWMA is a classificatory scheme that identifies progressive stages in the assimilated uses of workplace literacy. It reflects the perspective that an effective workplace participant is able to move through increasingly complex levels of thinking and to develop independence of thought and judgment (Knefelkamp, 1999). The profile of an individual who assimilates workplace norms can be characterized in five developmental stages:

1. *Concept recognition*: The first stage represents the capacity to learn, conceptualize, and articulate key issues related to the six sectors of workplace literacy. Concept recognition provides the basis for becoming adaptive to all workplace requirements.

2. *Multiple workplace perspectives*: This refers to the ability to integrate points of view from different colleagues at various levels of the workplace hierarchy. By using multiple perspectives, the individual is in a position to augment his or her workplace literacy.

3. *Comprehension of business processes*: Individuals increase their understanding of workplace cooperation, competition, and advancement as they build on their recognition of business concepts and workplace perspectives. They increasingly understand the organization as a system of interconnected parts.

4. *Workplace competence*: As assimilation and competence increase, the individual learns not only on how to perform a particular job adequately but how to conduct oneself professionally within the workplace and larger business environment.

5. *Professional independence*: Individuals demonstrate the ability to employ all sectors of workplace literacy to compete effectively in corporate labor markets. They obtain more responsible jobs through successful interviewing and workplace performance and demonstrate leadership abilities, leading to greater independence in career pursuits. Professionally independent individuals are motivated and can use their skills for creative purposes (Langer, 2009).

The LWMA is a rubric that charts an individual's development across the six sectors of workplace literacy. Each cell within the matrix represents a particular stage of development relative to that sector of workplace literacy, and each cell contains definitions that can be used to identify where a particular individual stands in his or her development of workplace literacy.

STAGES OF WORKPLACE LITERACY					
SECTORS OF WORKPLACE LITERACY	CONCEPT RECOGNITION	MULTIPLE WORKPLACE PERSPECTIVES	COMPREHENSION OF BUSINESS PROCESSES	WORKPLACE COMPETENCE	PROFESSIONAL INDEPENDENCE
Cognition					
Technology					
Business Culture					
Socio-Economic Values					
Community and Ethnic Solidarity					
Self-Esteem					

The LWMA and Action Research

While the LWMA serves as a framework for measuring growth, the model also uses reflection-with-action methods, a component of action research theory, as the primary vehicle for assisting young adults to develop the necessary labor market skills to compete for a job and inevitably achieve some level of professional independence (that is, the ability to work for many employers because of achieving required market skills). Reflection-with-action is used as a rubric for a variety of methods, involving reflection in relation to learning activities. Reflection has received a number of definitions from different sources in the literature. Here, "reflection-with-action" carries the resonance of Schön's (1983) twin constructs: "reflection-on-action" and "reflection-in-action," which emphasize (respectively) reflection in retrospect and reflection to determine what actions to take in the present or immediate future (Langer, 2003). Dewey (1933) and Hullfish and Smith (1978) also suggest that the use of reflection supports an implied purpose. Their formulation suggests the possibility of reflection that is future oriented; what we might call "reflection-to-action." These are methodological orientations covered by the rubric.

Reflection-with-action is critical to the educational and workplace assimilation process of Gen Y. While many people reflect, it is in being reflective that people bring about "an orientation to their everyday lives" (Moon, 2000). The LWMA incorporates reflection-with-action methods as fundamental strategies for facilitating development and assimilation. These methods are also implemented interactively, for example in mentoring, reflective learning journals, and group discussions. Indeed, as stated by De Jong (2014), "Social exclusion is multi-dimensional, ranging from unemployment, barriers to education and health care, and marginalized living circumstances" (p. 94). Ultimately, teaching socially excluded youth to reflect-with-action is the practice that will help them mature across the LWMA stages and inevitably, achieve levels of inclusion in the labor market and in citizenship.

Implications for New Pathways for Digital Talent

The salient implications of the LWMA, as a method of discovering and managing disadvantaged Gen Y youth in communities, can be categorized across three frames: demographic shifts in talent resources, economic sustainability and integration and trust among vested local interest groups.

Demographic Shifts in Talent Resources

The LWMA can be used as a predictive analytic tool for capturing and cultivating the abilities in the new generation of digital natives from disadvantaged local communities. This young talent has the advantage of more exposure to technologies, which senior workers had to learn later in their careers. This puts them ahead of the curve with respect to basic digital skills. Having the capacity to employ talent locally and provide incentives for these individuals to advance can alleviate the significant strain placed on firms who suffer from high turnover in outsourced positions. Investing in viable Gen Y underserved youth can help firms close the skills gap that is prevalent in the emerging labor force.

Economic Sustainability

As globalization ebbs and flows, cities need to establish themselves as global centers, careful not to slip into market obsolescence, especially when facing difficulties in labor force supply chains. In order to alleviate the difficulty in supplying industry-ready professionals to a city only recently maturing into the IT-centric business world, firms need to adapt to an "on-demand" gig approach. The value drawn from this paradigm lies in its cyclical nature. By obtaining *localized* human capital at a lower cost, firms can generate a fundable supply chain of talent and diversity as markets change over time.

Integration and Trust

Porter and Kramer (2011) postulate that companies need to formulate a new method of integrating business profits and societal responsibilities. They state, "the solution lies in the principle of shared value, which involves creating economic value in a way that also creates value for society by addressing its needs and challenges" (p. 64). Porter and Kramer suggest that companies need to alter corporate performance to include social progress. The LWMA provides the mechanism, theory, and measurement that is consistent with this direction and provides the vehicle that establishes a shared partnership of trust among business, education, and community needs. Each of the interested parties experiences progress toward its financial and social objectives. Specifically, companies are able to attract diverse and socially excluded local talent, and have the constituents trained specifically for its needs and for an economic return that fits its corporate models. As a result, the community adds jobs, which reduces crime rates and increases tax revenue. The funding corporation then establishes an ecosystem that provides a shared value of performance that underserved and excluded youth bring to the business.

Global Implications for Sources of Talent

The increasing social exclusion of Gen Y youth is a growing problem in almost every country. Questions remain about how to establish systemic solutions that can create sustainable and scalable programs that provide equity in access to education for this population. This access to education is undoubtedly increasing employability, which indirectly contributes to better citizenship for underserved youth. Indeed, there is a widening gap between the "haves" and the "have-nots" throughout the world. Firms can use tools like the LWMA to provide a model that can improve educational attainment of underserved youth by establishing skill-based certificates with universities, coupled with a different employment-to-hire model. The results have shown that students accelerate in these types of programs and ultimately, find more success in labor market assimilation. The data suggests that traditional degree programs that require full-time study at university as the primary preparation for labor market employment may not be the most appropriate approach to solving the growing social inequality issue among youth.

Conclusion

This chapter has made the argument that Gen Y employees are "digital natives" who have the attributes to assist companies to transform their workforce and meet the accelerated change in the competitive landscape. Organizations today need to adapt their staff to operate under the auspices of ROD by creating processes that can determine the strategic value of new emerging technologies and establish a culture that is more "change ready." Most executives across industries recognize that digital technologies are the most powerful variable to maintaining and expanding company markets.

Gen Y employees provide a natural fit for dealing with emerging digital technologies. However, success with integrating Gen Y employees is contingent upon baby boomer and Gen X management adapting new leadership philosophies and procedures that are suited to meet the expectations and needs of millennials. Ignoring the unique needs of Gen Y employees will likely result in an incongruent organization that suffers high turnover of young employees who will ultimately seek a more entrepreneurial environment. Firms should consider investing in non-traditional Gen Y youth from underserved and socially excluded populations as alternate sources of talent.

12

TOWARD BEST PRACTICES

Introduction

The previous chapters provided the foundation for the formation of "best practices" to implement and sustain responsive organizational dynamism (ROD). First, it is important to define what we mean by best practices and specify which components comprise that definition. Best practices are defined as generally accepted ways of doing specific functions or processes by a particular profession or industry. Best practices, in the context of ROD, are a set of processes, behaviors, and organizational structures that tend to provide successful foundations to implement and sustain organizational learning. I defined responsive organizational dynamism as the disposition of a company to respond at the organizational level to the volatility of advancing technologies—ones that challenge the organization to manage a constant state of dynamic and unpredictable change. Second, best practices are those that need to be attributed to multiple communities of practice as well as to the different professions or disciplines within a learning organization.

However, these multiple tiers of best practices need to be integrated and to operate with one another to be considered under the rubric. Indeed, best practices contained solely within a discipline or community are limited in their ability to operate on an organization-wide level. It is the objective of this chapter, therefore, to formulate a set of distinctive yet integrated best practices that can establish and support ROD through organizational learning. Each component of the set of best practices needs to be accompanied by its own maturity arc, which defines and describes the stages of development and the dimensions that comprise best practices. Each stage defines a linear path of continued progress until a set of best practices is reached. In this way, organizations can assess where they are in terms of best practices and determine what they need to do to progress. Ultimately, each maturity arc will represent a subset of the overall set of best practices for the organization.

The discipline that lays the foundation for ROD is information technology (IT). Therefore, the role of the chief IT executive needs to be at the base of organizational best practices. As such, I start building the organizational best practices model with the chief IT executive at the core.

Chief IT Executive

I use the title "chief IT executive" to name the most senior IT individual in an organization. Because of the lack of best practices in this profession, a number of different titles are used to describe this job. While these titles are distinct among themselves, I have found that they are not consistently followed in organizations. However, it is important to understand these titles and their distinctions, particularly because an organizational learning practitioner will encounter them in practice. These titles and roles are listed and discussed next:

> *Chief information officer (CIO)*: This individual is usually the most senior IT executive in an organization, although not every organization has such a person. The CIO is not necessarily the most technical of people or even someone who has come through the "ranks" of IT. Instead, this individual is considered an executive who understands how technology needs to be integrated within the organization. CIOs typically have other general IT executives and managers who report directly to them. As shown in the Siemens case study, there can be a number of alternate levels of CIOs, from corporate CIOs to local CIOs of a company division. For the purposes of this discussion, I look at the corporate CIO, who is considered part of the senior executive management team. My research on chief executive officer (CEO) perceptions of technology and business strategy showed that only a small percentage of CIOs report directly to the CEO of their organization, so it would be incorrect to generalize that they report to the most senior executive. In most cases, the CIO reports to the chief operating officer (COO) or the chief financial officer (CFO). As

stated, the role of the CIO is to manage information so that it can be used for business needs and strategy. Technology, then, is considered a valuable part of knowledge management from a strategic perspective as opposed to just a technical one.

Chief technology officer (CTO): This individual, unlike the CIO, is very much a senior technical person. The role of the CTO is to ensure that the organization is using the best and most cost-effective technology to achieve its goals. One could argue that the CTO holds more of a research-and-development type of position. In many organizations, the CTO reports directly to the CIO and is seen as a component of the overall IT infrastructure. However, some companies, like Ravell and HTC, only have a CTO and view technology more from the technical perspective.

Chief knowledge officer (CKO) and chief digital officer (CDO): This role derives from library management organizations because of the relevance of the word *knowledge* and/or *data*. It also competes somewhat with the CIO's role when organizations view technology from a perspective that relates more to knowledge. In larger organizations, the CKO/CDO may report directly to the CIO. In its purest role, the CKO/CDO is responsible for developing an overall infrastructure for managing knowledge, including intellectual capital, sharing of information, and worker communication. Based on this description, the CKO/CDO is not necessarily associated with technology but is more often considered part of the technology infrastructure due to the relevance of knowledge and data to technology.

To define best practices for this function, it is necessary to understand the current information and statistics about what these people do and how they do it. Most of the statistical data about the roles and responsibilities of chief IT executives are reported under the auspices of the CIO. According to an article by Jerry Gregoire in *CIO* magazine in March 2002, 63% of IT executives held the title CIO, while 13% were CTOs; there were few to no specific statistics available on the title of CKO and CDO, however, the CDO role has become more relevant over the past five years given the importance of social media and digital transformations. This report further supported the claim that there is limited use of the CKO title and function in organizations at this time.

From a structural point of view, 63% of IT organizations are centrally structured, while 23% are decentralized with a central reporting structure. However, 14% are decentralized without any central headquarters or reporting structure. From a spending perspective, organizations spend most of their budgets on integrating technology into existing applications and daily processing (36% of budget). Twenty-six percent is related to investments in emerging or new technologies, 24% is based on investing in e-commerce activities, and 24% is spent on customer relationship management (CRM), which is defined as applications that engage in assisting organizations to better understand and support their customer base. Twenty-five percent is spent on staff development and retention.

Compensation of IT chief executives still comes predominantly from base salary, as opposed to bonus or equity positions with the company. This suggests that their role is not generally viewed as top management or partner-level in the business. This opinion was supported by the results of my CEO study, discussed in Chapter 2. The issue of executive seniority can be determined by whether the chief IT executive is corporate driven or business unit driven. This means that some executives have corporate-wide responsibilities as opposed to a specific area or business unit. The issue of where IT departments provide value to the organization was discussed in Chapter 3, which showed that there are indeed different ways to manage and structure the role of IT. However, in general, corporate IT executives are responsible for IT infrastructure, shared technology services, and global technology architecture, while business unit CIOs concentrate on strategically understanding how to use applications and processes to support their business units. This is graphically depicted in Figure 12.1.

From a best practices perspective, the following list has historically suggested what chief IT executives should be doing. The list emphasizes team building, coaching, motivating, and mentoring as techniques for implementing these best practices.

Strategic thinking: Must understand the business strategy and competitive landscape of the company to apply technology in the most valuable way to the organization.

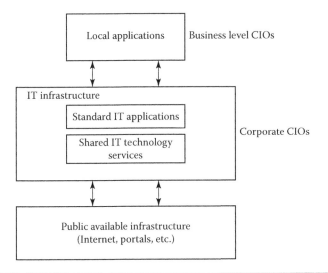

Figure 12.1 Business-level versus corporate-level CIOs.

Industry expertise: Must have the ability to understand the product and services that the company produces.

Create and manage change: Must have the ability to create change, through technology, in the operating and business processes of the organization to gain efficiency and competitive advantage.

Communications: Must have the ability to communicate ideas, to give direction, to listen, to negotiate, to persuade, and to resolve conflicts. Executives must also be able to translate technical information to those who are not technologically literate or are outside IT and need to be comfortable speaking in public forums and in front of other executives.

Relationship building: Must have the ability to interface with peers, superiors, and customers, by establishing and maintaining strong rapport, bond, and trust between individuals.

Business knowledge: Must have the ability to develop strong business acumen and having peripheral vision across all functional areas of the business.

Technology proficiency: Must have the knowledge to identify appropriate technologies that are the most pragmatic for the business, can be delivered quickly at the lowest cost, produce an impact on the bottom line (ROI), and have longevity.

Leadership: Must be a visionary person, inspirational, influential, creative, fair, and open minded with individuals within and outside the organization.

Management skills: Must have the ability to direct and supervise people, projects, resources, budget, and vendors.

Hiring and retention: Must have the ability to recognize, cultivate, and retain IT talent.

While this list is not exhaustive, it provides a general perspective, one that appears generic; that is, many management positions in an organization might contain similar requirements. A survey of 500 CIOs conducted by *CIO* magazine (March 2002) rated the top three most important concerns among this community in terms of importance:

1. Communications: 70%
2. Business understanding: 58%
3. Strategic thinking: 46%

What is interesting about this statistic is that only 10% of CIOs identified technical proficiency as critical for their jobs. This finding supports the notion that CIOs need to familiarize themselves with business issues, as opposed to just technical ones. Furthermore, the majority of a CIO's time today has been recorded as spent communicating with other business executives (33%) and managing IT staffs (28%). Other common activities reported in the survey were as follows:

- Operating the baseline infrastructure and applications
- Acting as technology visionary
- Implementing IT portions of new business initiatives
- Designing infrastructure and manage infrastructure projects
- Allocating technology resources
- Measuring and communicate results
- Serving as the company spokesperson on IT-related matters
- Selecting and managing product and service providers
- Recruiting, retaining, and developing IT staff
- Participating in company and business unit strategy development

These results further confirm that chief IT executives define best practices, based on understanding and supporting business strategy.

This survey also reported common barriers that chief IT executives have to being successful. The overarching barrier that most IT executives face is the constant struggle between the business expectation to drive change and improve processes, and the need to reduce costs and complete projects faster. The detailed list of reported problems by rank was as follows:

1. Lack of key staff, skill sets, and retention: 40%
2. Inadequate budgets and prioritizing: 37%
3. Shortage of time for strategic thinking: 31%
4. Volatile market conditions: 22%
5. Ineffective communications with users: 18%
6. Poor vendor support and service levels, and quality: 16%
7. Overwhelming pace of technological change: 14%
8. Disconnection with executive peers: 12%
9. Difficulty proving the value of IT: 10%
10. Counterproductive office politics: 6%

Chief IT executives also felt that their roles were ultimately influenced by two leading factors: (1) changes in the nature and capabilities of technology, and (2) changes in the business environment, including marketplace, competitive, and regulatory pressures. This can be graphically viewed in Figure 12.2.

Figure 12.2 has a striking similarity to Figure 3.1 outlining ROD. That diagram represented technology as an independent variable creating the need for ROD, which is composed of strategic integration and cultural assimilation, as shown in Figure 12.3.

Figure 12.3 shows many similarities to Figure 12.2. The difference between these two diagrams defines what is missing from many best practices: the inclusion of organizational learning practices that would enable chief IT executives to better manage business and technology issues. In effect, if organizational learning techniques were included, they could reduce many barriers between business and IT. Thus, the solution to providing best practices for the IT community rests with the inclusion of organizational learning along the constructs of ROD.

The inclusion of organizational learning is crucial because the best practices, as reported among the community of chief IT executives, has

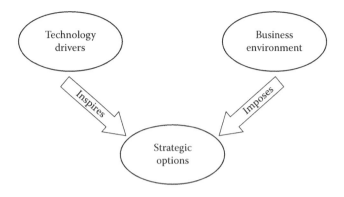

Figure 12.2   Chief IT executives—factors influencing strategic options.

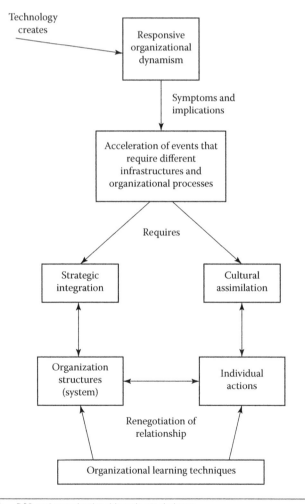

Figure 12.3   ROD and organizational learning techniques.

not produced the performance outcomes sought by chief executives. I refer to Chapter 2, in which I first defined the IT dilemma. While many IT initiatives are credible, they often fall short of including critical business issues. As a result, IT project goals are not completely attained. This suggests that the problem is more related to the process and details of how to better implement good ideas. As further support for this position, the Concours Group (an international executive managing consulting organization) published a list of emerging roles and responsibilities that chief IT executives will need to undertake as part of their jobs in the near future (Cash & Pearlson, 2004):

Shared services leader: More companies are moving to the shared services model for corporate staff functions. CIOs' experiences may be invaluable in developing and managing these organizations.

Executive account manager: More companies today are involving the CIO in the management of relationships between the company and its customers.

Process leader: As companies move toward organizing around major business processes, a CIO is in a good role to temporarily lead this effort since applications and databases are among the business resources that must be revamped to implement process management.

Innovation leader: A CIO is starting to act as the innovation leader of the corporation when a company is seeking to achieve substantial improvements in process performance or operational efficiencies, or to implement IT, since innovation may center on the application of IT.

Supply chain executive: Purchasing, warehousing, and transportation are among the most information-intensive activities undertaken by a business. As companies look to improve these overall processes, the CIO may become the most knowledgeable executive about the supply chain.

Information architect: Companies are recognizing the benefit of a consolidated view of customers, vendors, employees, and so on. CIOs are finding themselves taking on the leadership role of information architect by cultivating commitment and consensus around this challenging task.

Change leader: CIOs are playing an increasingly important role in business change management. Their role is in either direct change leadership (developing new business models) or, more often, indirect; that is, change process is behind the scenes (get other leaders to think about new possibilities).

Business process outsourcing leader: CIOs tend to have some of the most extensive experience in company outsourcing. This makes them a logical internal consultant and management practice leader in business process outsourcing.

These issues all suggest that the role of the chief IT executive is growing and that the need for these executives to become better integrated with the rest of their organizations is crucial for their success. Much more relevant, though, is the need for ROD and the role that the chief IT executive has as a member of the overall community. To create best practices that embrace organizational learning and foster ROD, a chief IT executive maturity arc needs to be developed that includes the industry best practices presented here integrated with organizational learning components.

The chief IT executive best practices arc is an instrument for assessing the business maturity of chief IT executives. The arc may evaluate a chief IT executive's business leadership using a grid that measures competencies ranging from essential knowledge in technology to more complex uses of technology in critical business thinking. Thus, the chief IT executive best practices arc provides executives with a method of integrating technology knowledge and business by presenting a structured approach of self-assessment and defined milestones.

The model measures five principal facets of a technology executive: cognitive, organization culture, management values, business ethics, and executive presence. Each dimension or sector is measured in five stages of maturation that guide the chief IT executive's growth. The first facet calls for *becoming reflectively aware* about one's existing knowledge of technology and what it can do for the organization. The second calls for *other centeredness*, in which chief IT executives become aware of the multiplicity of technology perspectives available (e.g., other business views of how technology can benefit the organization). The third is *comprehension of the technology process*, in which a chief IT executive can begin to merge technology issues with business concepts and functions.

The fourth is *stable technology integration*, meaning that the chief IT executive understands how technology can be used and is resilient to nonauthentic sources of business knowledge. Stage 4 represents an ongoing implementation of both technology and business concepts. The fifth stage is *technology leadership*, in which chief IT executives have reached a stage at which their judgment on using technology and business is independent and can be used to self-educate from within. Thus, as chief IT executives grow in knowledge of technology and business, they can become increasingly more other centered, integrated, stable, and autonomous with the way they use their business minds and express their executive leadership and character.

Definitions of Maturity Stages and Dimension Variables in the Chief IT Executive Best Practices Arc

Maturity Stages

1. *Technology competence and recognition*: This first stage represents the chief IT executive's capacity to learn, conceptualize, and articulate key issues relating to cognitive technological skills, organization culture/etiquette, management value systems, business ethics, and executive presence needed to be a successful chief IT executive in business.

2. *Multiplicity of technology perspectives*: This stage indicates the chief IT executive's ability to integrate multiple points of view about technology from others in various levels of workplace hierarchies. Using these new perspectives, the chief IT executive augments his or her skills with the technology necessary for career success, expands his or her management value system, is increasingly motivated to act ethically, and enhances his or her executive presence.

3. *Comprehension of technology process*: Maturing chief IT executives accumulate increased understanding of workplace cooperation, competition, and advancement as they gain new cognitive skills about technology and a facility with business culture/etiquette, expand their management value system, perform business/workplace actions to improve ethics about business and technology, and develop effective levels of executive presence.

4. *Stable technology integration*: Chief IT executives achieve integration with the business community when they have levels of cognitive and technological ability, organization etiquette/culture, management values, business ethics, and executive presence appropriate for performing job duties not only adequately but also competitively with peers and even higher-ranking executives in the workplace hierarchy.

5. *Technology leadership*: Leadership is attained by the chief IT executive when he or she can employ cognitive and technological skills, organization etiquette, management, a sense of business ethics, and a sense of executive presence to compete effectively for executive positions. This chief IT executive is capable of obtaining increasingly executive-level positions through successful communication and workplace performance.

Performance Dimensions

1. *Technology cognition*: Concerns skills specifically related to learning, applying, and creating resources in IT, which include the necessary knowledge of complex operations. This dimension essentially establishes the CIO as technically proficient and forms a basis for movement to more complex and mature stages of development.

2. *Organizational culture*: The knowledge and practice of proper etiquette in organizational settings with regard to dress, telephone and in-person interactions, punctuality, work completion, conflict resolution, deference, and other protocols in workplace hierarchies.

3. *Management values*: Measures the individual's ability to articulate and act on mainstream organizational values credited with shaping the work ethic—independent initiative, dedication, honesty, and personal identification with career goals, based on the organization's philosophy of management protocol.

4. *Business ethics*: Reflects the individual's commitment to the education and professional advancement of other employees in technology.

Developmental dimensions of maturing

Dimension skill	Technology competence and recognition	Multiplicity of technology perspectives	Comprehension of technology process	Stable technology integration	Technology leadership
Technology cognition					
Organization culture					
Management values					
Business ethics					
Executive presence					

Figure 12.4   Chief IT executive best practices arc - conditions for assessment.

5. *Executive presence*: Involves the chief IT executive's view of the role of an executive in business and the capacity to succeed in tandem with other executives. Aspects include a devotion to learning and self-improvement, self-evaluation, the ability to acknowledge and resolve business conflicts, and resilience when faced with personal and professional challenges.

Figure 12.4 shows a graphical view of the chief IT executive best practices arc. Each cell in the arc provides the condition for assessment. The complete arc is provided in Table 12.1.

Chief Executive Officer

When attempting to define CEO best practices, one is challenged with the myriad material that attempts to determine the broad, yet important, role of the CEO. As with most best practices, they are typically based on trends and percentages of what most CEOs do—assuming, of course, that the companies they work for are successful. That is, if their organization is successful, then their practices must be as well. This type of associative thinking leads to what scholars often term false generalizations. Indeed, these types of inadequate methods lead to false judgments that foster business trends, which are misinterpreted as best practices. Reputation is what would better define these trends, which usually after a period of time can become ineffective

Table 12.1   The Chief IT Executive Best Practices Arc

DIMENSION VARIABLE	TECHNOLOGY COMPETENCE AND RECOGNITION	MULTIPLICITY OF TECHNOLOGY PERSPECTIVES
Technology Cognition	Understands how technology operates in business. Has mastered how systems are developed, hardware interfaces, and the software development life cycle. Has mastery of hardware, compilers, run-time systems. Has core competencies in distributed processing, database development, object-oriented component architecture, and project management. Is competent with main platform operating systems such as UNIX, WINDOWS, and MAC. Has the core ability to relate technology concepts to other business experiences. Can also make decisions about what technology is best suited for a particular project and organization. Can be taught how to expand the use of technology and can apply it to other business situations.	Understands that technology can have multiple perspectives. Able to analyze what are valid vs. invalid opinions about business uses of technology. Can create objective ideas from multiple technology views without getting stuck on individual biases. An ability to identify and draw upon multiple perspectives available from business sources about technology. Developing a discriminating ability with respect to choices available. Realistic and objective judgment, as demonstrated by the applicability of the technology material drawn for a particular project or task and tied to functional/ pragmatic results.
Organization Culture	Understands that technology can be viewed by other organizations in different ways. Uses technology as a medium of communication. Understands that certain technological solutions, Web pages, and training methods may not fit all business needs and preferences of the business. Has the ability to recommend/suggest technological solutions to suite other business needs and preferences	Seeks to use technology as a vehicle to learn more about organization cultures and mindsets. Strives to care about what others are communicating and embraces these opinions. Tries to understand and respect technologies that differ from own. Understands basic technological needs of others.

(*Continued*)

Table 12.1 (Continued) The Chief IT Executive Best Practices Arc

COMPREHENSION OF TECHNOLOGY PROCESS	STABLE TECHNOLOGY INTEGRATION	TECHNOLOGY LEADERSHIP
Has the ability to relate various technical concepts and organize them with non-technical business issues. Can operate with both automated and manual business solutions. Can use technology to expand reasoning, logic, and what-if scenarios. Ability to use the logic of computer programs to integrate the elements of non-technological tasks and business problems. Ability to discern the templates that technology has to offer in order to approach everyday business problems. This involves the hypothetical (inductive/deductive) logical business skill.	Knowledge of technology is concrete, accurate, and precise, broad and resistant to interference from non-authentic business sources. Ability to resist or recover from proposed technology that is not realistic—and can recover resiliently.	Methods and judgment in a multidimensional business world is independent, critical discernment. Knowledge of technology and skills in technology can be transferred and can be used to self-educate within and outside of technology. Can use technology for creative purposes to solve business challenges and integrate with executive management views.
Can deal with multiple dimensions of criticism about technology. Can develop relationships (cooperative) that are dynamic and based on written communication and oral discourse. Ability to create business relations outside of technology departments. Has an appreciation of cyberspace as a communication space—a place wide open to dialogue (spontaneous), to give and take, or other than voyeuristic, one-sidedness. Ability to produce in teamwork situations, rather than solely in isolation.	Loyalty and fidelity to relations in multiple organizations. Commitment to criticism and acceptance of multiple levels of distance and local business relationships. Ability to sustain non-traditional types of inputs from multiple sources.	Can utilize and integrate multidimensions of business solutions in a self-reliant way. Developing alone if necessary using other technical resources. Can dynamically select types of interdependent and dependent organizational relationships. Ability to operate within multiple dimensions of business cultures, which may demand self-reliance, independence of initiative, and interactive communications.

(Continued)

Table 12.1 (Continued)   The Chief IT Executive Best Practices Arc

DIMENSION VARIABLE	TECHNOLOGY COMPETENCE AND RECOGNITION	MULTIPLICITY OF TECHNOLOGY PERSPECTIVES
Management Values	Technology and cultural sensitivity. Global communication, education, and workplace use of technology can be problematic—subject to false generalizations and preconceived notions. Awareness of assumptions about how technology will be viewed by other organizations and about biases about types of technology (MAC vs. PC).	Can appreciate need to obtain multiple sources of information and opinion. The acceptance of multi-dimensional values in human character.
Business Ethics	Using technology with honesty re: privacy of access and information. Development of ethical policies governing business uses of the Internet, research, intellectual property rights and plagiarism.	The use of information in a fair way—comparison of facts against equal sources of business information. Compassion for business information for which sources are limited because of inequality of technology access. Compassion for sharing information with other business units from a sense of inequality.
Executive Presence	Has accurate perception of one's own potential and capabilities in relation to technology in the business—the technologically realizable executive self.	Understands how other executives can view self from virtual and multiple perspectives. Understands or has awareness of the construction of self that occurs in business. Focuses on views of other executives in multiple settings. Understands that the self (through technology) is open for more fluid constructions, able to incorporate diverse views in multiple settings.

(*Continued*)

Table 12.1 (Continued) The Chief IT Executive Best Practices Arc

COMPREHENSION OF TECHNOLOGY PROCESS	STAB LE TECHNOLOGY INTEGRATION	TECHNOLOGY LEADERSHIP
Can operate within multiple dimensions of value systems and can prioritize multi-tasking events that are consistent with value priorities. Ability to assign value to new and diverse technology alternatives—integrating them within a system of pre-existing business and technology values.	Testing value systems in new ways due to technology is integrated with long-term values and goals for business achievement. Some concepts are naturally persistent and endure despite new arenas in the technological era.	Use of technology and business are based on formed principles as opposed to dynamic influences or impulses. Formed principles establish the basis for navigating through, or negotiating the diversity of business influences and impulses.
Consistent values displayed on multiple business communications, deliverables of content, and dedication to authenticity. Maintains consistency in integrating values within technology business issues.	Technology is a commitment in all aspects of value systems, including agility in managing multiple business commitments. Commitment to greater openness of mind to altering traditional and non-technological methods.	Technological creativity with self-defined principles and beliefs. Risk-taking in technology-based ventures. Utilizing technology to expand one's arenas of business freedom. Exploring the business-liberating capacities of technology.
Operationalizes technology to unify multiple components of the self and understands its appropriate behaviors in varying executive situations.	Has regulated an identity of self from a multiplicity of executive venues. Methods of business interaction creates positive value systems that generate confidence about operating in multiple business communities.	Acceptance and belief in a multidimensional business world of the self. Can determine comfortably the authenticity of other executives and their view of the self. Can confirm disposition independently from others' valuations, both internally and from other organization cultures. Beliefs direct and control multidimensional executive growth.

and unpopular. We must also remember the human element of success; certain individuals succeed based on natural instincts and talent, hard work and drive, and so on. These components of success should not be confused with theories that are scalable and replicable to practice; that is, what best practices need to accomplish.

This section focuses on technology best practices of the CEO. These best practices are based on my research as well as other positions and facts that provide a defendable context of how and why they appear to be effective. However, as with the chief IT executive model, best practices cannot be attained without an arc that integrates mature organizational learning and developmental theories. Many of the CEO best practices reconcile with my interviews with CEOs and, in particular, with the two CEO case studies (of ICAP and HTC) discussed in Chapter 8. Other published definitions and support are referenced in my presentation.

In February 2002, Hackett Benchmarking, a part of Answerthink Corporation, issued its best practices for IT. Its documentation stated:

> In compiling its 2002 best practices trend data, Hackett evaluated the effectiveness (quality and value) and efficiency (cost and productivity) of the information technology function across five performance dimensions: strategic alignment with the business; ability to partner with internal and external customers; use of technology; organization; and processes.*

The findings, as they apply to the CEO function, provide the following generalizations:

- There was an 85% increase in the number of CIOs who reported directly to the CEO. This increase would suggest that CEOs need to directly manage the CIO function because of its importance to business strategy.
- CEOs supporting outsourcing did not receive the cost-cutting results they had hoped for. In fact, most broke even. This suggests that CEOs should not view outsourcing as a cost-cutting measure, but rather foster its use if there are identifiable business benefits.

---

* Hackett Benchmarking has tracked the performance of nearly 2,000 complex, global organizations and identified key differentiators between world-class and average companies, across a diverse set of industries. In addition to information technology, staff functions studied include finance, human resources, procurement, and strategic decision making, among others. Study participants comprised 80% of the Dow Jones Industrials, two-thirds of the Fortune 100, and 60% of the Dow Jones Global Titans Index. Among the IT study participants are Agilent Technologies, Alcoa, Capital One Financial Corporation, Honeywell International, Metropolitan Life Insurance, SAP America, and TRW. (From PR Newswire, February 2002.)

- CEOs have found that IT organizations that have centralized operations save more money, have fewer help-line calls than decentralized organizations, and do not sacrifice service quality. This suggests that the CEOs should consider less business-specific support structures, especially when they conduct their business at multiple locations.
- CEOs are increasingly depending on the CIO for advice on business improvements using technology. As a result, their view is that IT professionals need advanced business degrees.
- CEOs should know that consistent use of IT standards has enabled firms to trim IT development costs by 41%, which has reduced costs for end-user support and training operations by 17%.
- CEOs need to increase support for risk management. Only 77% of average companies maintained disaster recovery plans.

As we can see from these generalizations, they are essentially based on what CEOs are doing, and what they have experienced. Unfortunately, this survey addressed little about what CEOs know and exactly what their role should be with respect to overall management, participation, and learning of technology. These "best practices" are particularly lacking in the area of organizational learning and the abilities of the firm to respond to changing conditions as opposed to searching for general solutions. Let us look at each of these generalizations and discuss what they lack in terms of organizational learning.

CIO Direct Reporting to the CEO

The fact that more CIOs are reporting directly to the CEO shows an escalation of their importance. But, what is more relevant as a best practice is what that relationship is about. Some report about how often they meet. What is more important is the content of the interactions. What should the CEOs know, how should the CEOs conduct themselves? What management and learning techniques do they apply? How do they measure results? My CEO interview research exposed the fact that many CEOs simply did not know what they needed to do to better manage the CIO and what they needed to know in general about technology.

Outsourcing

Outsourcing can be a tricky endeavor. In Chapter 3, I introduced the concept of technology as a driver and a supporter. I presented a model that shows how emerging technologies are initially drivers and need to be evaluated and measured using similar models embraced by marketing-oriented communities. I then showed how, through maturation, emerging technologies become supporters, behaving more as a commodity within the organization. I explained that only then can a technology be considered for outsourcing because supporter operations are measured by their economies of scale, reduced costs, increased productivity, or both (efficiency). Figure 12.5 shows that cycle.

Thus, what is missing from the survey information is the knowledge of where such technologies were with respect to this technology life cycle. Knowing this dramatically affects what the CEO should be expecting and what organizational learning concepts and factors are needed to maximize benefit to the organization.

Centralization versus Decentralization of IT

The entire question of how IT should or should not be organized must be based on a business that implements ROD. ROD includes the component called cultural assimilation, which provides a process,

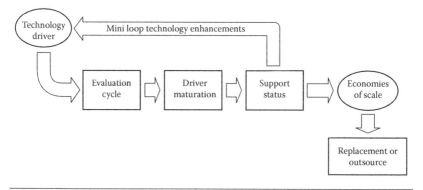

Figure 12.5   Driver-to-supporter life cycle.

using organizational learning, to help businesses determine the best IT structure. To simply assume that centralizing operations saves money is far too narrow; the organization may need to spend more money on a specific technology in the short term to get long-term benefits. Where is this included in the best practices formula? My research has shown that more mature uses of technology in organizations require more decentralization of IT personnel within the business units. The later stages of IT organizational structure at Ravell supported this position.

CIO Needs Advanced Degrees

I am not sure that anyone could ever disagree with the value of advanced degrees. Nevertheless, the survey failed to provide content on what type of degree would be most appropriate. It also neglected to address the issue of what may need to be learned at the undergraduate level. Finally, what forms of education should be provided on the job? What exactly are the shortfalls that CIOs need to know about business? And, equally important is the consideration of what education and learning is needed by CEOs and whether they should be so dependent on advice from their CIOs.

Need for Standards

The need for standards is something that most organizations would support. Yet, the Siemens case study showed us that too much control and standardization can prove ineffective. The Siemens model allowed local organizations to use technology that was specific to their business as long as it could be supported locally. The real challenge is to have CEOs who understand the complexity of IT standards. They also need to be cognizant that standards might be limited to the structure of their specific organization structure, its business, and its geographical locations.

Risk Management

The survey suggested that CEOs need to support risk management because their backup recovery procedures may be inadequate. The question is whether the problem stems from a lack of support or from a lack of knowledge about the topic. Is this something that the chief IT executive needs to know, or is it just about the CEO's unwillingness to spend enough funds? The best practices component of risk management must be broader and answer these questions.

By contrast to the survey, we may consider a report issued by Darwin Research ("A CRM Success Story," November 1, 2002), which cited the recommended best practices of Christopher Milliken, CEO of Boise Cascade Office Products. He offered the kind of in-depth view of best practices that I feel is needed to be consistent with my research on ROD. Milliken participated in the implementation of a large-scale CRM system needed to give his customers a good reason to choose Boise. The project required an investment of more than $20 million. Its objective was to provide customers with better service. At the time of the investment, Milliken had no idea what his ROI would be, only that the project was necessary to distinguish Boise Cascade from myriad competitors in the same industry.

After the successful implementation of the project, Milliken was now in a position to offer his own thoughts about technology-related best practices that a CEO might want to consider. He came up with these six:

1. *The CEO must commit to a technology project*: Milliken was keen to express the reasons why the CRM project was important; he was intimately involved with its design, and made it clear that he had to be consulted, should there be any delays in the project schedule. KPMG (a major consulting firm) was also hired as a consultant to help implement the schedule and was held to the same level of excellence. What Milliken accomplished, significantly, was to show his interest in the project and his willingness to stay involved at the executive level. Milliken's best practice here lies in his commitment, which is consistent with that of McDermott from ICAP and the CEO from HTC. They both realized, as Milliken did, that the CEO must have an active role in the project and not just allow the management team to get it done. Milliken, as did McDermott and the CEO of HTC, issued specific performance-related requirements to his employees and consultants.

His participation sent a valuable message: The CEO is part of the supporting effort for the project and is also part of the learning process of the organization. Indeed, the situation that Milliken faced and resolved (i.e., to jump in without knowing the expected returns of the project) is exemplary of the core tenets of ROD, which require the ability for an organization to operate with dynamic and unpredictable change brought about by technology. In this case, the technology was crucial to distinguishing Boise Cascade, in the same way that electronic trading was for ICAP, and the billing system was for HTC. Yet, all three of these situations required a certain behavior and practice from the company CEO. Thus, the most important best practice lies in the commitment and learning to the learning organization format.

2. *Think business first, then technology*: To understand why a technology is needed, there must first be a supporting business plan; that is, the business plan must drive the technology or support its use. This best practice concept is consistent with my research. Indeed, Dana Deasy from Siemens realized it after a three-year investment in e-business, and McDermott clearly advocated the importance of a business plan over embellished technology. Another interesting and important result of the business plan was that it called for the creation of a centralized CRM system. Therefore, it became necessary to consolidate the separate business units at Boise into one corporate entity—providing central support and focus. This is another example of how ROD operates. The CRM project, through a validation process in a business plan, provided the strategic integration component of ROD. The strategy then influenced cultural assimilation and required a reorganization to implement the strategy or the new CRM system. Furthermore, Boise Cascade allowed its staff to experiment in the project, to make mistakes, without criticizing them. They were, in effect, implementing the driver-related concepts of technology. These driver concepts must be similar to the way organizations support their marketing activities, by which they accept a higher error ratio than when implementing a supporter activity. The CEO wanted everyone to give it their

best and to learn from the experience. This position is a key best practice for the CEO; it promotes organizational learning throughout the business.

3. *Handcuff business and technology leaders to each other*: Milliken understood that technology projects often fail because of a lack of communication between IT and other business entities. The project represented many of the IT dilemmas that I discussed in Chapter 2, particularly relating to the new CRM system and its integration with existing legacy applications and, at the same time, creating a culture that could implement the business strategy. To address this, Milliken first appointed a new CIO to foster better communication. He also selected a joint project leader from the business side, thus creating a joint project leadership team. What Milliken did was to form a new community of practice that did not exist before the project. The project, as with Ravell, represented an event that fostered the creation of organizational learning opportunities. As with ICAP, Milliken's company enlisted the support of executive-level consultants to help finalize the business plan and marketing strategy, as well as assist with change management. What exactly did Milliken do that represents a best practice? From an organizational learning perspective, he created communities of practice between IT and the business. That then is a true best practice for a CEO.

4. *Get the show on the road*: There was a not-to-be-questioned deadline that was instituted by Milliken. As I noted in Chapter 4, this type of management seems undemocratic, but it should not be confused with being nonparticipatory. Someone had to get this going and set expectations. In this case, both IT and business users were set to make things happen. Senior management endorsed the project and openly stated that it represented what could be a one-time opportunity to "do something of great magnitude" (Dragoon, 2002). From a best practices perspective, this means that the CEO can and should provide the leadership to get projects done and that part of that leadership could be setting strategic dates. However, CEOs should not confuse this leadership with power-centralized management over IT-related projects.

Communities of practice still need to be the driving force for inevitable success in ROD. Another important factor was Milliken's decision to create dual management over the project. Thus, Milliken was able to create an environment that required discourse between IT and the business.

5. *Win over the masses for massive changes*: As stated, the business plan called for a reorganization of other business units. This also required executives to rethink job descriptions and titles in relation to new processes. It also eliminated six redundant management-level jobs. Milliken engaged employees in a massive "external-internal" marketing campaign. Employees participated in ad campaigns, and brochures were created for all staff. A video was also produced that defined the benefits to Boise Cascade customers. In essence, Milliken was committed to communication and training. Similar to my experience at Ravell, not everyone is comfortable with change, and resistance in the ranks is inevitable. As a result, the education and training programs at Boise were not enough. What was lacking was true organizational learning and knowledge management. There are two best practices that were defined from this experience. First, the CEO needs to engage in actively showing the importance that technology has to the organization, not only from an economic perspective, but also from a staff development point of view. The second best practice comes from the example of what Boise Cascade did not do enough of: provide organizational transformation through knowledge management, reflective practices, and communities of practice. This suggests that CEOs need to better understand and incorporate organizational learning concepts, so that they can be the catalyst for change as they are in other areas of the business. We saw support for this concept from both ICAP and HTC, where the actions of the CEO came from an organizational learning perspective.

6. *Know that technology projects never end*: ROD assumes, by definition, that technology is a variable, albeit an insistent one. Milliken's experience further supported this notion, in that he realized that Boise Cascade must continue to assess the impact of the CRM application. Another way of saying this

is that the technology will continue to be viewed as a means to transform the business on an ongoing basis. Indeed, Milliken was planning to spend another $10 million on the next phase. So, from a best practices perspective, CEOs must recognize that technology investment never ends, but it moves to other phases of maturation, similar to the driver/supporter life cycle. Finally, the buy-in to this reality ensures the recognition of organizational dynamism.

Based on the case studies and research presented thus far in this book, I can now formulate a list of 11 key planks that represent the core of what constitutes a technology CEO's set of best practices:

1. The chief IT executive should report directly to the CEO.
2. CEOs should be actively committed to technology on an ongoing basis, as opposed to project-by-project involvement.
3. CEOs should be willing to be management catalysts to support new technology-driven projects. They, in effect, need to sometimes play the role of technology champion.
4. CEOs should focus on business concepts and plans to drive technology. In other words, technology should not drive the business.
5. CEOs should use consultants to provide objective input to emerging technology projects.
6. CEOs should establish organizational infrastructures that foster the creation of communities of practice. They need to create joint ownership of IT issues by fostering discourse between IT, business managers, and staff.
7. CEOs may need to take control of certain aspects of technology investments, such as setting milestones and holding management and staff to making critical project dates.
8. CEOs need to foster cultural assimilation, which may lead to reorganization, since technology changes processes.
9. CEOs need to understand organizational learning and knowledge management theories and participate in organizational transformation.
10. CEOs need to understand how the technology life cycle behaves, with specific attention to the transition from driver activities to supporter functions. To that end, CEOs need to

understand the short- and long-term investments that need to be made in technology.

11. CEOs should create organizations that can effectively operate within technological dynamism. This process will educate management and staff to handle the dynamic and unpredictable effects of emerging technologies. It will also foster the development of both middle-up-down and bottom-up management of technology.

The issue is now to provide a linear development model for CEOs that enables them to measure where they are in relation to ROD and the best practices outlined.

The CEO Best Practices Technology Arc

Similar to the chief IT executive arc, the CEO best practices arc is an instrument for assessing the technology best practices of CEOs. The arc evaluates a CEO's strategic uses of technology and leadership by using a grid that charts competencies ranging from conceptual knowledge about technology to more complex uses of technology and business and how they are integrated in strategic business planning.

As with all arc models, the CEO version measures five principal stages of a CEO's maturity with respect to business applications of technology: conceptual, structural, executive values, executive ethics, and executive leadership. Each dimension or sector is measured in five stages of maturation that guide the CEO's executive growth managing technological dynamism. The first stage is being reflectively aware about their conceptual knowledge of technology and what it can do for the organization. The second is other centeredness, by which CEOs become aware of the multiplicity of business uses of technology and the different views that can exist inside and outside the organization. The third is integration of business use of technology; a CEO can begin to combine how business plans foster the need for technology. The fourth is implementation of business/technology process, meaning that the CEO understands how business applications and technology are used together and is resilient to nonauthentic sources of emerging technologies. Stage four represents an ongoing commitment to implementing both technology and business applications. The fifth refers to strategic uses of technology; CEOs have reached a stage at which their judgment on using technology and business is independent and can be used to self-educate. Thus, as CEOs grow in knowledge of business uses of technology, they can become increasingly more understanding of the multiplicity of uses, can become more integrated in how they conceptualize technology, can manage its implementation from an executive position, and can apply strategies to support new applications of technology in the organization.

Definitions of Maturity Stages and Dimension Variables in the CEO Technology Best Practices Arc

Maturity Stages

1. *Conceptual knowledge of technology*: This first stage represents the CEO's capacity to learn, conceptualize, and articulate key issues relating to business uses of technology, organizational structures available, executive value methods, executive ethical issues surrounding technology, and leadership alternatives that are needed to be successful with technology applications.

2. *Multiplicity of business perspectives of technology*: This stage indicates the CEO's ability to integrate multiple points of view from management, staff, and consultants about technology applications in business. Using these new perspectives, the CEO augments his or her conceptual skills with technology, has an expanded view of what organizational structures might work best, expands his or her executive values about technology uses, is increasingly aware of the ethical dilemmas with technology, and enhances his or her leadership abilities.

3. *Integration of business uses of technology*: Maturing CEOs accumulate increased understanding of how technology can support the business, provide more competitive advantage, and have a more integrated understanding of how to use their conceptual skills about technology, of the alternative organizational structures available, of how to combine their business executive value and ethical systems, and how to develop effective levels of executive leadership.

4. *Implementation of business/technology process*: CEOs achieve integration when they can regularly apply their conceptual knowledge of technology, organization structures, executive values and ethics about technology, and executive leadership, appropriate for performing their job duties, not only adequately, but at a level that provides a competitive advantage for the organization.

5. *Strategic uses of technology*: Leadership is attained by the CEO when he or she can employ conceptual skills, develop new organizational structures as necessary, establish new values and ethics that are appropriate for the organization, and create a sense of executive presence to lead the organization strategically. This CEO is capable of having new vision about how business and technology can be expanded into new endeavors.

Performance Dimensions

1. *Technology concepts*: Concerns conceptual skills, specifically related to understanding how technology can be used in the business. This dimension essentially establishes the CEO as technically proficient, conceptually, and forms a basis for movement to more complex and mature stages of business/technology development.

2. *Organizational structures*: The knowledge of the alternative organizational structures that can support the application of emerging technology in corporate settings with regard to roles, responsibilities, career paths, and organizational reporting alternatives.

3. *Executive values*: Measures the CEO's ability to articulate and act on mainstream technological values credited with shaping the work ethic: independent initiative, dedication, honesty, and personal identification with career goals, based on the philosophy of the management protocol of the organization.

4. *Executive ethics*: Reflects the CEO's commitment to the education and professional advancement of the behavior of the organization as it relates to business uses of technology.

5. *Executive leadership*: Involves the CEO's view of the role of an executive in business, and the capacity to succeed in tandem with his or her organizational resources. Aspects include a devotion to organizational learning and self-improvement, self-evaluation, the ability to acknowledge and resolve business/technology conflicts, and resilience when faced with personal and professional challenges.

Figure 12.6 shows a graphic view of the CEO technology best practices arc. Each cell in the arc provides the condition for assessment. The complete arc is provided in Table 12.2.

Middle Management

Middle management, which comprises a number of tiers, is perhaps the most challenging of best practices to define. In Chapter 3, I stratified the different types of positions that make up middle managers into three tiers: directors, line managers, and supervisors. What is most important at this point is to determine the set of technology best practices for managers so that they can effectively operate under ROD. That is, technology best practices must be designed to contain the insights and skills for effective management of technology. This must include

1. Working with IT personnel
2. Providing valuable input to the executive management team, including the CEO
3. Participating and developing a technology strategy within their business units
4. Effectively managing project resources, including technical staff
5. Leading innovative groups in their departments
6. Incorporating technology into new products and services
7. Developing proactive methods of dealing with changes in technology
8. Investigating how technology can improve competitive advantage.

Developmental dimensions of maturing

Dimension skill	Conceptual knowledge of technology	Multiplicity of business perspectives of technology	Integration of business uses of technology	Implementation of business/technology process	Strategic uses of technology
Technology cognition					
Organizational structures					
Executives values					
Executives ethics					
Executive leadership					

Figure 12.6   CEO technology best practices arc.

Table 12.2   CEO Technology Best Practices Arc—Detail

DIMENSION VARIABLE	CONCEPTUAL KNOWLEDGE OF TECHNOLOGY	MULTIPLICITY OF BUSINESS PERSPECTIVES OF TECHNOLOGY
Technology Concept	Understands concepts and definitions about technology and how it relates to business. Has conceptual knowledge of the software development life cycle. Understands high-level concepts about distributed processing, database development, and project management. Understands the definition and role of operating systems such as UNIX, WINDOWS, and MAC. Has the ability to relate technology concepts to other business experiences. Understands that different technology may be required for a particular project and organization. Can conceptualize how to expand the use of technology and apply it to business situations.	Seeks to manage by appreciating that technology can have multiple perspectives. Able to manage a process that requires validation about different opinions about business uses of technology. Can manage the different objective ideas from multiple technology views without getting stuck on personal biases. Has an ability to identify and draw upon multiple perspectives available from business sources about technology, particularly from independent sources. Developing a discriminating ability to create an infrastructure that can operate with multiple views. Committed to creating an organization that can learn through realistic and objective judgment, as demonstrated by the applicability of the technology material drawn for a particular project or task and tied to business outcomes.
Organizational Structures	Understands that technology can be viewed by other organizations in different ways and may need different organizational structures. Can use technology as a medium of communication. Understands that certain technologies may need to be managed differently and need specific types of structures and expertise. Has the ability to comprehend recommend/suggested technological solutions to suite business needs and preferences.	Seeks to manage technology as a vehicle to learn more about what alternative organization structures are available from others. Strives to create a learning organization that cares about what other staff perceive as solutions. Committed to cultural assimilation that can change the need to restructure the organization. Tries to understand and respect technologies that differ from what the organization is currently using. Understands that the organization has multiple and different technological needs.

(*Continued*)

Table 12.2 (Continued)   CEO Technology Best Practices Arc—Detail

INTEGRATION OF BUSINESS USES OF TECHNOLOGY	IMPLEMENTATION OF BUSINESS/TECHNOLOGY PROCESS	STRATEGIC USES OF TECHNOLOGY
Creates an organization that has the ability to relate various technical concepts and organize them with non-technical business issues. Can manage by operating with both automated and manual business solutions. Can use technology to expand business reasoning, logic, and what-if scenarios. Establishes business templates that allow technology to offer everyday business solutions. This involves the hypothetical (inductive/deductive) logical business issues.	Organization's use of technology is concrete, accurate, and precise, broad and resistant to interference from non-authentic technology business sources. Ability to resist or recover from faulty uses of technology that is not realistic without a supporting business plan.	Methods and judgment as a multidimensional CEO is independent, has critical discernment. Conceptual knowledge of technology can be transferred and can be used to self-educate within and outside of technology. Can use technology for creative purposes to create new business initiatives and integrate them with short- and long-term business goals.
Can deal with multiple dimensions of criticism about how technology can be used in the organization. Can develop relationships (cooperative) that are dynamic and based on written communication and oral discourse about how business can drive technological investments. Ability to create new business relations using technology with new and existing customers. Has an appreciation of cyberspace as a new market—a place wide open to dialogue (spontaneous), to provide new opportunities for business growth.	Commitment to open discussion of alternating opinions on technology and acceptance of varying types of structures to accommodate technology opportunities. Ability to sustain dynamic organizational structures.	Can design new structures to integrate multidimensions of business and technology solutions. Can dynamically manage different types of interdependent and dependent organizational relationships. Ability to manage within multiple dimensions of business cultures, which may demand self-reliance and confidence in independence of initiatives.

(Continued)

Table 12.2 (Continued)   CEO Technology Best Practices Arc—Detail

DIMENSION VARIABLE	CONCEPTUAL KNOWLEDGE OF TECHNOLOGY	MULTIPLICITY OF BUSINESS PERSPECTIVES OF TECHNOLOGY
Executive Values	Understanding of technology and cultural differences. Conceptually understands that global communication, education, and workplace use of technology can be problematic—subject to false generalizations and preconceived notions. Management awareness of responsibilities to address assumptions about how technology will be viewed by other departments and customers.	Sets conditions that foster the need to obtain multiple sources of information and opinion about how technology values. The propagation organizationally of acceptance that there can be multidimensional values in human character.
Executive Ethics	Understands that there is a need to use technology with honesty re: privacy of access and information. Supports the development of ethical policies governing business uses of the Internet, research, intellectual property rights, and plagiarism.	Committed to creating an organization that uses information in a fair way—comparison of facts against equal sources of business information. Understands and is compassionate that business and technology information may have different levels of knowledge access. Recognizes the need for sharing information with other business units from a sense of inequality.
Executive Leadership	Conceptualizes the need to have a leadership role with respect to technology in the business—the business and technologically realizable executive self.	Understands how other executives can view technology leadership differently. Understands or has awareness of the construction of self that occurs when taking on the integration of technology in business operations. Focuses on views of other CEOs in multiple settings. Understands that the self (through technology) is open for more fluid constructions, able to incorporate diverse views in multiple technology settings.

(Continued)

Table 12.2 (Continued)   CEO Technology Best Practices Arc—Detail

INTEGRATION OF BUSINESS USES OF TECHNOLOGY	IMPLEMENTATI ON OF BUSINESS/TECHNOLOGY PROCESS	STRAT EGIC USES OF TECHNOLOGY
Can manage multiple dimensions of value systems and can prioritize multi-tasking events that are consistent with value priorities. Ability to assign value to new and diverse technology business alternatives—linking them to legacy systems and processes.	Managing value systems in new ways because technology changes long-term values and goals for business objectives. Recognition that some concepts remain unchanged despite emerging technologies.	Management of technology and business are based on formed principles as opposed to dynamic influences or impulses. Formed executive principles establish the basis for navigating through or negotiating the diversity of business opportunities and impulses for investment in technologies.
Consistent management values displayed on multiple business goals, mission, and dedication to authenticity. Maintains management consistency in combining values regarding technology issues.	Business and technology are a commitment in all aspects of management value systems, including agility in managing multiple business commitments. Commitment to greater openness of mind to altering traditional and non-technological management methods.	Technology management creativity with self-defined principles and beliefs. Risk-taking in technology-based ventures. Utilizing technology to expand one's arenas of business development. Manages the business liberating capacities of technology.
Manages technology to unify multiple parts of the organization and understands how the process behaves in different business situations.	Has developed an executive identity of self from a multiplicity of management venues. Method of management creates positive value systems that generate confidence about how multiple business communities need to operate.	Acceptance and belief in a multidimensional business world of how to lead with technology. Can determine comfortably, authenticity of organization's executives and their view of the self. Can confirm disposition on technology independently from others' valuations, both internally and from other organizations. Beliefs direct and control multi-dimensional leadership growth.

As with CEO research, there are myriad best practices that have been offered as a method of dealing with the subject of technology management. Unfortunately, these practices usually are vague and intermingle management levels and departments; that is, it is difficult to know whether the best practice is for the chief IT executive, the CEO, or some other level of management. We know from the research from Bolman and Deal (1997) that middle managers feel torn by conflicting signals and pressures they get from both senior management and the operations that report to them: "They need to understand the difference in taking risks and getting punished for mistakes" (p. 27). According to Bolman and Deal (1997), best practices for middle managers need to cover the following areas:

1. Knowledge management
2. Alignment
3. Leadership and commitment
4. Organization
5. Human resources
6. Opportunity management
7. Leveraging
8. Performance assessment

Their study covered more than 400 companies in the eight areas of concern. I extracted 10 middle management-related best practices from their study results and concluded that middle managers need to

1. Understand how to take a strategy and implement it with technology; that is, they need to create tactics for completing the project.
2. Establish team-building measures for linking technology with daily operations of the staff.
3. Foster the aggregation and collaboration of business unit assets to form peer groups that can determine joint efforts for implementing new technologies.
4. Stimulate their staffs using innovative strategies of value propositions and reward systems.
5. Create multifunctional teams that can focus on particular aspects of how technology affects their specific area of expertise.
6. Follow common project management practices so that multitier and department projects can be globally reviewed by senior management.
7. Form project teams that can respect and perform on an action basis; that is, teams that are action oriented.

8. Understand how to communicate with, and use, IT staff on projects.
9. Have a systematic process for gathering intelligence relating to pertinent technology developments.
10. Understand that customers are the drivers for technology tools provided by the organization.

On reviewing the different aspects of middle manager best practices with technology research, it appears that there are two focal points: (1) those best practices that address the needs of senior management, the CIO, and the CEO; and (2) those that are geared toward the management of the staffs who need to implement emerging technology projects.

This makes sense, given that the middle manager, notwithstanding whether a director, line manager, or supervisor, needs to deal with executive productivity-related issues as well as staff implementation ones. They are, as Bolman and Deal (1997) state, "torn" by these two competing organizational requirements. Table 12.3 represents the combined list of technology-based best practices organized by executive best practices and implementation best practices.

Table 12.3 exemplifies the challenge that middle managers have in balancing their priorities. In accordance with the research, the best practices mentioned are implemented using methods of knowledge management, alignment, leadership and commitment, human resources, opportunity management, leveraging, and performance assessment. As with the other best practices, the middle manager technology best practices are limited because they do not address the specific needs of ROD, particularly organizational learning theories (with the exception of knowledge management). This shortfall is integrated into another developmental arc model that combines these theories with the preceding definitions of best practices.

The Middle Management Best Practices Technology Arc

The middle management best practices technology arc, as with others, can be used to evaluate a middle manager's strategic and operational

Table 12.3   Middle Manager Executive and Implementation Best Practices

EXECUTIVE-BASED MIDDLE MANAGER BEST PRACTICES	IMPLEMENTATION-BASED MIDDLE MANAGER BEST PRACTICES
1. Provide valuable input to the executive management team, including the CEO.	1. Understand how to communicate with and use IT staff on projects.
2. Incorporate technology into new products and services.	2. Effectively manage project resources, including technical staff.
3. Participate in developing a technology strategy within their business units.	3. Lead innovative groups in their departments.
4. Have proactive methods of dealing with changes in technology.	4. Understand how to take a strategy and implement it with technology; that is, create tactics for completing the project.
5. Focus on how technology can improve competitive advantage.	5. Establish team-building measures for linking technology with staff's daily operations.
6. Have a systematic process for gathering intelligence, relating to pertinent technology developments.	6. Foster the aggregation and collaboration of business unit assets to form peer groups that can determine joint efforts for implementing new technologies.
7. Understand that customers are the drivers for technology tools provided by the organization reward.	7. Stimulate their staffs using innovative strategies of value propositions and systems.
	8. Create multifunctional teams that can focus on particular aspects of how technology affects their specific area of expertise.
	9. Follow common project management practices so that multitier and department projects can be globally reviewed by senior management.
	10. Form project teams that can respect and perform on an action basis; that is, teams that are action oriented.

uses of technology by using a grid that measures competencies ranging from conceptual knowledge about technology to more complex uses of technology and business operations.

The five principal stages defined by the arc determine the middle manager's maturity with business implementations of technology: cognitive, organization interactions, management values, project ethics, and management presence. There are five stages of maturation that guide the middle manager's growth. The first is becoming reflectively aware about one's existing knowledge with business technology and how it can be implemented. The second is the recognition of the multiplicity of ways that technology can be implemented on projects (e.g., other business views of how technology can benefit the organization). The third is integration of business implementation with technology, in which a middle manager can begin to combine technology issues with business concepts and functions on a project basis. The fourth is stability of business/technology implementation, in which the middle manager has integrated business/technology as a regular part of project implementations. The fifth is technology project leadership, in which the middle manager can use their independent judgment on how best to use technology and business on a project-by-project basis. Thus, as middle managers grow in knowledge of technology and business projects, they can become increasingly more open to new methods of implementation and eventually, autonomous with the way they implement projects and provide leadership.

Definitions of Maturity Stages and Dimension Variables in the Middle Manager Best Practices Arc

Maturity Stages

1. *Technology implementation competence and recognition:* This first stage represents the middle manager's capacity to learn, conceptualize, and articulate key issues relating to cognitive business technological skills, organizational interactions, management value systems, project management ethics, and management presence.

2. *Multiplicity of business implementation of technology:* Indicates the middle manager's ability to integrate multiple points of view during technical project implementations. Using these new perspectives, the middle manager augments his or her skills with business implementation with technology career advancement, expands his or her management value system, is increasingly motivated to act ethically during projects, and enhances his or her management presence.

3. *Integration of business implementation of technology:* Maturing middle managers accumulate increased understanding of how business and technology operate together and affect one another. They gain new cognitive skills about

technology and a facility with how the organization needs to interact, expand their management value system, perform business/technology actions to improve ethics about business and technology, and develop effective levels of management presence.

4. *Stability of business/technology implementation:* Middle managers achieve stable integration when they implement projects using their cognitive and technological ability; have organization interactions with operations; have management values with their superiors, peers, and subordinates; possess project ethics; and have the management presence appropriate for performing job duties, not only adequately, but also competitively (with peers and higher-ranking executives in the organization hierarchy).

5. *Technology project leadership:* Leadership is attained by the middle manager when he or she can employ cognitive and technological skills, organization interactions, management, a sense of business ethics, and a sense of management presence to compete effectively for executive positions. This middle manager is capable of obtaining increasingly executive-level positions through successful interviewing and organization performance.

Performance Dimensions

1. *Business technology cognition*: Pertains to skills specifically related to learning, applying, and creating resources in business and technology, which include the necessary knowledge of complex operations. This dimension essentially establishes the middle manager as "operationally" proficient with technology and forms a basis for movement to more complex and mature stages of development when managing technology projects.

2. *Organizational interactions*: This focuses on the middle manager's knowledge and practice of proper relationships and management interactions during technology projects. This pertains to in-person interactions, punctuality of staff, work

completion, conflict resolution, deference, and other protocols in technology projects.

3. *Management values*: Measures the middle manager's ability to articulate and act on mainstream corporate values credited with shaping technology project work ethic: independent initiative, dedication, honesty, and personal identification with technology project goals, based on the philosophy of management protocol of the organization.

4. *Project ethics:* Reflects the middle manager's commitment to the education and professional advancement of other persons in technology and in other departments.

5. *Management presence:* Involves the middle manager's view of the role of a project-based manager during a technology project implementation and the capacity to succeed in tandem with other projects. Aspects include a devotion to learning and self-improvement, self-evaluation, the ability to acknowledge and resolve business conflicts, and resilience when faced with personal and professional challenges during technology implementations.

Figure 12.7 shows a graphic view of the middle management technology best practices arc. Each cell in the arc provides the condition for assessment. The complete arc is provided in Table 12.4. The challenge of the middle management best practices arc is whether to emphasize executive management concepts (more organizationally intended) or event-driven concepts (project oriented). This arc focuses on project implementation factors and deals with best practices that can balance executive pressures with implementation realities. I suggest that senior middle managers, at the director level, who do not participate in implementation, set their best practices, based on the CEO maturity arc. Indeed, creating a separate arc for upper management would contain too many overlapping cells.

Summary

The formation of best practices to implement and sustain ROD is a complex task. It involves combining traditional best practice methods (i.e., what seems to work for proven organizations and individuals)

Developmental dimensions of maturing

Dimension skill	Technology implementation	Multiplicity of business	Integration of business	Stability of business/technology	Technology project
Business technology					
Organizational					
Management values					
Project ethics					
Management presence					

Figure 12.7   Middle management technology best practices arc.

Table 12.4   Middle Management Technology Best Practices Arc—Detail

DIMENSION VARIABLE	TECHNOLOGY IMPLEMENTATION COMPETENCE AND RECOGNITION	MULTIPLICITY OF BUSINESS IMPLEMENTATION OF TECHNOLOGY
Business Technology Cognition	Understands how technology operates during projects. Has conceptual knowledge about hardware interfaces, and the software development life cycle. Has the core ability to relate technology concepts to other business experiences. Can also participate in the decisions about what technology is best suited for a particular project. Can be taught how to expand the use of technology and can apply it to other business situations.	Understands that technology projects can have multiple perspectives on how to implement them. Able to analyze what is valid vs. invalid opinions about business uses of technology. Can create objective ideas from multiple technology views without getting stuck on individual biases. An ability to identify and draw upon multiple perspectives available from project sources about technology. Developing a discriminating ability with respect to choices available. Realistic and objective judgment, as demonstrated by the applicability of the technology material drawn for a particular project or task and tied to functional/pragmatic outcomes.
Organizational Interactions	Understands that technology projects require the opinions of other departments and staff in multiple ways. Understands that certain technological solutions and training methods may not fit all project needs and preferences of the business. Has the ability to recommend/suggest alternative technological solutions to suite other business and project needs and preferences.	Seeks to use technology projects as a vehicle to learn more about organization interactions and mindsets. Strives to care about what others are communicating and embraces these opinions on a project basis. Tries to understand and respect technologies that differ from own. Understands basic technological project needs of others.

(Continued)

Table 12.4 (Continued)   Middle Management Technology Best Practices Arc—Detail

INTEGRATION OF BUSINESS IMPLEMENTATION OF TECHNOLOGY	STABILITY OF BUSINESS/ TECHNOLOGY IMPLEMENTATION	TECHNOLOGY PROJECT LEADERSHIP
Has the ability to relate various technical project concepts and organize them with non-technical business issues. Can operate with both business and technical solutions. Can use technology to expand reasoning, logic, and what-if scenarios. Ability to discern the templates that technology has to offer in order to approach everyday technology project problems. This involves the hypothetical (inductive/deductive) logical business and technology skills.	Knowledge of technology projects are concrete, accurate, and precise, broad and resistant to interference from non-authentic business and technical project sources. Ability to resist or recover from proposed technology that is not realistic—and can recover resiliently.	Methods and judgment in multidimensional technology projects are independent and use critical discernment. Operational knowledge of technology and project management skills can be transferred and can be used to self-educate within and outside of technology. Can use technology for creative purposes to solve business and project challenges and integrate with executive management views.
Can deal with multiple dimensions of criticism about technology-based projects. Can develop relationships (cooperative) that are dynamic and based on discourse. Ability to create project relations with IT, other departments, and customers. Has an appreciation of project communication—to foster open dialogue (spontaneous), to give and take, or other than voyeuristic, one-sidedness about the project. Ability to produce in teamwork situations, rather than solely in isolation.	Loyalty and fidelity to relations in multiple organizations. Commitment to criticism and acceptance of multiple levels of IT and business relationships. Ability to sustain non-traditional types of inputs from multiple sources during projects.	Can utilize and integrate multidimensions of project solutions in a self-reliant way. Developing alone if necessary using other technical and non-technical resources. Can dynamically select types of interdependent and dependent organizational relationships. Ability to operate within multiple dimensions of business cultures, which may demand self-reliance, independence of initiative, and interactive communications during project implementations.

(Continued)

Table 12.4 (Continued)   Middle Management Technology Best Practices Arc—Detail

DIMENSION VARIABLE	TECHNOLOGY IMPLEMENTATION COMPETENCE AND RECOGNITION	MULTIPLICITY OF BUSINESS IMPLEMENTATION OF TECHNOLOGY
Management Values	Technology and cultural sensitivity during project implementations. Global communication, education, and project use of technology can be problematic—subject to false generalizations and preconceived notions. Awareness of assumptions about how technology will be viewed by other departments and staff and about biases about types of technology used (MAC vs. PC).	Can appreciate need to obtain multiple sources of information and opinions during project implementations. The acceptance of multidimensional values in human character as value during project design and completion.
Project Ethics	Using technology on the project with honesty re: privacy of access and information. Development of ethical policies governing project uses of the Internet, research, intellectual property rights, and plagiarism.	The use of information in a fair way—comparison of facts against equal sources of project information. Compassion for differences in project information for which sources are limited because of inequality of technology access. Compassion for sharing information with other business units from a sense of inequality.
Management Presence	Has accurate perception of one's own potential and capabilities in relation to technology projects—the technologically realizable manager.	Understands how other managers can view self from a virtual and multiple perspectives. Understands or has awareness of the construction of self that occurs in projects. Understands views of other executives and managers in multiple project settings. Understands that the self (thru technology projects) are open for more fluid constructions, able to incorporate diverse views in multiple settings.

(*Continued*)

Table 12.4 (Continued)   Middle Management Technology Best Practices Arc—Detail

INTEGRATION OF BUSINESS IMPLEMENTATION OF TECHNOLOGY	STABILITY OF BUSINESS/ TECHNOLOGY IMPLEMENTATION	TECHNOLOGY PROJECT LEADERSHIP
Can operate project within multiple dimensions of value systems and can prioritize multitasking events that are consistent with value priorities. Ability to assign value to new and diverse technology project alternatives—integrating them within a system of pre-existing business and technology project implementation values.	Testing technology value systems in new ways during the project implementation is integrated with long-term values and goals for business achievement. Some project concepts are naturally persistent and endure despite new arenas in the technological era	Use of technology and business during project implementation are based on formed principles as opposed to dynamic influences or impulses. Formed principles establish the basis for navigating through, or negotiating the diversity of business influences and impulses during the project.
Consistent values displayed on multiple project communications, deliverables of content, and dedication to authenticity. Maintains consistency in integrating values within technology business issues during project implementation.	Technology is a commitment in all aspects of value systems, including agility in managing multiple project commitments. Commitment to greater openness of mind to altering traditional and non-technological methods on project implementations.	Technological project creativity with self-defined principles and beliefs. Risk-taking in technology-based projects. Utilizing technology to expand one's arenas of project freedom. Exploring the project management liberating capacities of technology.
Operationalizes technology projects to unify multiple components of the self and understands its appropriate behaviors in varying management situations.	Has regulated an identity of self from a multiplicity of management venues. Method of project interaction creates positive value systems that generate confidence about operating in multiple organizational communities.	Can determine comfortably, authenticity of other managers and their view of the self. Can confirm project-related disposition independently from others' valuations, both internally and from other department cultures. Has direct beliefs and controls multidimensional management growth.

with developmental theory on individual maturation. The combination of these two components provides the missing organizational learning piece that supports the attainment of ROD. Another way of comprehending this concept is to view the ROD arc as the overarching or top-level model. The other maturity arcs and best practices represent the major communities of practice that are the subsets of that model. This is graphically depicted in Table 12.5.

Thus, the challenge is to create and sustain each community and, at the same time, establish synergies that allow them to operate together. This is the organizational climate created at ICAP, where the executive board, senior and middle managers, and operations personnel all formed their own subcommunities; at the same time, all had the ability for both downward and upward communication. In summary, this particular model relies on key management interfaces that are needed to support ROD.

Ethics and Maturity

The word *ethics* is defined in many different ways. Reynolds (2007) defines ethics as "a set of rules that establishes the boundaries of generally accepted behaviour" (p. 3). Ethics can also mean conforming to social norms and rules, which can be challenged by deviant behaviors of "others." Still other groups construct ethics as a moral code that a community agrees to uphold. Ethics often map to our values—like integrity and loyalty to others. What is ethical for one person may not be ethical for another. This issue frames yet another question: How does ethics relate to leadership, specifically leadership in technology?

Ethics became a heightened issue after the Enron scandal in the United States. The scandal had a huge effect on the IT industry because it resulted in Congress enacting the Sarbanes-Oxley (SOX) Act, which placed significant audit trail requirements on documenting processes. Most of these processes existed in automated applications; thus, IT was required to comply with the rules and regulations that the SOX Act mandated. Implementing the SOX Act became an immense challenge for IT organizations mostly because the rules of compliance were vague.

Most would agree that ethics are a critical attribute for any leader. The challenge is how to teach it. The SOX Act "teaches" ethics by establishing governance by control—control of unethical behavior through catching deviants. However, history has shown us that deviants are not cured by laws and punishment; rather, they are simply contained. Unfortunately, containment does not eliminate or cure unethical behavior. Furthermore, deviants tend to find new ways to

Table 12.5 ROD and Best Practices Arcs

UNDERLINING BEST PRACTICES	UNDERLINING BEST PRACTICES	UNDERLINING BEST PRACTICES
Strategic-inking	Committed to technology	MANAGEMENT-BASED
Industry expertise	Technology catalyst and champion	Interact with executive management
Change management	Business first, then technology	Incorporate technology into new products
Communications	Use consultants for objective input	Use technology for competitive advantage
Business knowledge	Support communities of practice	Process for evaluating new technologies
Technology proficiency	Set project milestones	Understand driver role of customers
Hiring and retention	Foster cultural assimilation	
Innovation and outsourcing leadership	Understand organizational learning	IMPLEMENTATION-BASED
Information architect	Understand technology life cycle	Utilization of it staff on projects
	Have chief it exec report directly	Leading innovative groups
	Support organizational dynamism	Effectively managing project resources
		Strategic use of technology
		Establish team-building measures
		Foster aggregation and collaboration
		Stimulate staff with value propositions
		Create multi-functional teams
		Support common project management practices
		Form action-oriented teams

bypass controls and get around the system in time. On the other hand, educators more often see the solution as transforming behavior of the individual; that is, ethics can only be taught if the individual realizes its value. Value in ethical behavior becomes a systemic transformation when the individual believes in its self-realization. Being ethical is then aligned with self-actualization and adult maturity. So, ethics can be aligned with maturity in the same ways that the maturity arcs presented were mapped to leadership. Why is this so important for IT leaders? The SOX Act answered this question because it clearly identified the IT function as the most critical component of compliance. Unethical behavior in technology-based systems can damage the greater good, which places a big responsibility on the IT function.

I would suggest that IT ethics and leadership are very much linked. It is a very important responsibility for technology executives to provide direction to their firms on how technology and ethics are integrated and how they can transform individuals to value conformance without the overuse of governing controls. Firms must use organizational learning tools as the vehicle to promote such conformance through changes in behavior. Unfortunately, many executives, including those in IT, practice governance much more than influence. I am not suggesting the elimination of controls, but rather, that leadership should depend less on governance and more on effecting behavioral change. In other words, the key to developing strong ethics within an IT organization is leadership, not governance. An important component of leadership is the ability to influence the behaviors of others (without exerting control or power). The real power of leadership is to use influence to effect ethical behavior as opposed to demanding it.

How do we create ethical IT organizations? Further, how can a technology executive provide the necessary strategy and influence to accomplish firm-wide ethical transformation? The first strategy, for a number of reasons, should be to create an ethical IT organization as the model:

1. The technology executive has control over that organization.
2. Most IT ethical problems today emanate from technology personnel because of their unusual access to data and information.
3. IT is positioned to lead the direction, since it is its area of expertise.

So, IT can set the example for technology-related ethics for the entire organization by establishing its own level of compliance by a "way of being," as opposed to a way of being managed. Often, this way of being can evolve into a code of behavior that can become the cultural "code" of the organization itself. This code of ethics should address and be limited to such IT-related issues as:

- *Privacy*: Because of their access to transactions over the Internet, IT professionals must respect the privacy of information of others. Their code of ethics should go beyond just e-mail transactions to include access to personal data that may be stored on desktops or data files.
- *Confidentiality*: This differs from privacy because the data are available to IT in the normal transactions of business. That is, the data are captured or used in the development of an application. IT personnel need to keep such information confidential at all times—not only for the employees of the firm but also for clients and vendors.
- *Moral responsibility*: IT needs to protect the organization from outside abuses or questionable transactions coming into and leaving the company. Protection can also include blocking access to certain websites that are dangerous or inappropriate. This practice should not be regarded as a control, but rather, as a moral responsibility of any employee. Of course, there needs to be careful objectivity in how the moral code is actually executed when a problem is identified.
- *Theft*: Removing information that belongs to someone else can be construed as a form of theft. Theft should always be regarded as an offense punishable by law—that is, above and beyond rules and regulations of the company.

These are only examples of areas in which an ethical code might be applied. Such a code must be implemented in IT as a framework for how people are employed and as a basis for promotion. Again, governance plays an important part because unfortunately there will always be individuals who violate ethics. What we need are organizations that promote and defend ethics to the greatest possible extent. This way of being is consistent with the core definition of a learning organization in that ethics must inevitably be part of the fabric of the culture and evolved within it. With IT serving as a model, the technology executive can act as the champion for implementation company-wide. This chapter has shown that ethics are intrinsically linked to maturity. Indeed, every arc contained a dimension that contained an ethical dimension. Perhaps if such ethical practices existed at Enron, the "learning organization" there could have stopped the abuses.

13

CONCLUSION

Introduction

This book has explored many conceptual aspects of information technology (IT) and organizational learning and how they can be utilized together to help firms compete in a rapidly changing world. Case studies were presented to show how these concepts, and the theories they derive from, could be implemented into practice. It is most important, however, to remember that each organization is unique and that the implementation of organizational learning methods must therefore be tailored to the particular dynamics at play in a given organization. Hence, there can be no boilerplate methodology for the strategic employment of technology; such an approach could never guarantee maximum benefit to the organization. My position involves employing various organizational learning methods that must be carefully chosen and implemented, based on the projected target audience and on the particular stage of growth of the organization and its mature use of technology.

In my study of chief executive officer (CEO) perceptions of IT, I found that the role of IT was not generally understood in most of the organizations I surveyed, especially at the CEO level. There appear to be inconsistent reporting structures within the IT organization, and there is a lack of IT-related discussion at the strategic and senior executive levels. Furthermore, most executives are not satisfied with IT performance, and while most agree that technology should play a larger role in marketing, few have been able to accomplish this. The general dilemma has involved an inability to integrate technology effectively into the workplace.

Certainly, a principle target of this book is to answer the question of what chief IT executives need to do and in what directions their roles need to evolve regarding IT. Other concerns center on general organizational issues surrounding who IT people are, where they report, and how they should be evaluated. IT must also provide better leadership with respect to guiding a company through the challenges of unproven technologies. While technology behaves dynamically, we still need processes that can validate its applicability to the organization. Another way of viewing this is to accept the idea that certain technologies need to be rejected because of their inappropriateness to drive strategy.

IT is unique in that it is often viewed from a project perspective; for instance, that which is required to deliver technology and the cultural impact it has on the organization, and tends to be measured by project deliverables due to the pressure to see measurable outcomes. From a project perspective, IT staff members typically take on the role of project managers, which requires them to communicate with multiple business units and management layers. They need to establish shorter project life cycles and respond to sudden changes to the requirements. No longer does a traditional project life cycle with static dates and deliverables work with the fast-paced businesses of today. Rather, these projects are living and breathing entities that must be in balance with what is occurring in the business at all times. Most important is that project measurable outcomes must be defined and seen in balance with expected organizational transformations.

I began my explanation of the role of technology by establishing it as a dynamic variable, which I termed technological dynamism. Responsive organizational dynamism (ROD) represents my attempt to think through a range of responses to the problems posed by technological dynamism, which is an environment of dynamic and unpredictable change resulting from the advent of innovative technologies. This change can no longer be managed by a group of executives or managers; it is simply too complex, affecting every component of a business. A unilateral approach does not work; the problem requires an environmental approach. The question is how to create an organization that can respond to the variability of technologies in such a way that its responses become part of its everyday language and discourse. This technological state of affairs is urgent for two major reasons. First, technology not only is an accelerator of change but also requires accelerated business responses. Organizations cannot wait for a committee to be formed or long bureaucratic processes to act. Second, the market is unforgiving when it comes to missing business opportunities. Every opportunity missed, due to lack of responding in a timely fashion, can cost an organization its livelihood and future. As stated by Johansen et al. (1995):

> The global marketplace requires constant product innovation, quick delivery to market, and a large number of choices for the consumer, all of which are forcing us to rethink the way we structure our business organizations to compete. Indeed, many businesses are finding their traditional structure cumbersome—the way they work is more of an obstacle than help in taking advantage of global opportunities. (p. 1)

While ROD is the overarching approach for a firm that can perform in a dynamic and unpredictable environment, there are two major components to that approach that I raised for further consideration. I discussed how technology, as a variable, is unique in that it affects two areas of any organization. The first is the technology itself and how it operates with business strategy. I called this the strategic integration component of responsive organizational dynamism. The challenge here is to have organizations create processes that can formally and informally determine the benefit of new and emerging technologies on an ongoing basis. The second component is cultural assimilation, which is about managing the cultural and structural changes that are required when new strategies are adopted by the organization.

Creating an environment of ROD requires processes that can foster individual and organizational-level thinking, learning, and transformation. Organizational learning techniques best fit the need as they contain the core capabilities to assist organizations in reinventing themselves as necessary, and to build an organization that can evolve with technology, as opposed to one that needs to be reorganized. I have presented many organizational learning concepts and modified them to provide specific remedies to the challenges required to create responsive organizational dynamism. I have also presented the complex vectors that determine which learning theory should be applied and integrated with others, so that every aspect of individual and organizational evolution can be supported. I chose to use the term *vector* to describe this force or influence because of the different ways in which these learning methods can help in creating and sustaining firm-wide responses to technological dynamism.

Perhaps the most important learning process among these is that of linear development leading to maturation. My use of maturity arcs permits me a framework for the development and integration of models that can measure where individuals and organizations are in their trajectory toward the integration of emerging technologies in their business strategies. These maturity arcs provide a basis for how to measure where the organization is, what types of organizational learning methods to consider, and what outcomes to expect. Indeed, providing measurable outcomes in the form of strategic performance is the very reason why executives should consider embracing this model.

I also discussed a number of methods to manage organizational learning, modifying theories of knowledge management and change management, so that they specifically addressed the unique aspects of change brought about by new technologies. I looked at how the CEO needs to become more knowledgeable about technology, and, based on case studies and research, I provided sets of best practices to suggest that staff members cannot become part of a learning organization without the participation of the CEO and his or her executive committees. On the other hand, I investigated the interesting work of Nonaka and Takeuchi (1995) and their middle-up-down theory of middle management. I modified Nonaka and Takeuchi's idea by complicating the strata that can be used to define the middle, and I established three tiers of middle management and integrated them into organizational learning theories. Finally, I used the Ravell case study to show how operations personnel continue to play an important role in organizational learning, and how the maturity arc can be used to transform individual learning practices into less event-driven learning at the organizational level. I formulated best practices for each of these three major organizational structures, along with corresponding maturity arcs to lay the foundation of what each community needs to do to properly participate in the transformations indicated for responsive organizational dynamism. To this end, I proposed certain road maps that, if followed, could provide the mechanisms that lead to the kind of organizational transformation that is empowered to handle the challenges of new technologies. This process is summarized in Figure 13.1.

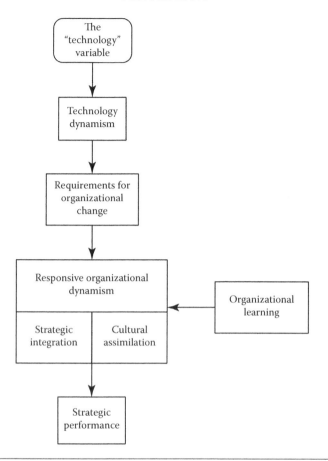

Figure 13.1 Technology "road map."

I have taken a strong position regarding the debate over whether learning occurs best on the individual level or at the system-organizational one, particularly as learning affects the establishing and sustaining of responsive organizational dynamism. My response to this debate is "yes"—yes, in the sense that both are very much needed and part of a process that leads to a structured way of maturing the use of organizational learning by an organization to improve strategic performance. I believe the Ravell case study provides an example of how learning maturation operates in a dynamic environment. We see that operations personnel tend to rely on event-driven and individual-based reflective practices before being able to think at an organizational level. My prior research (Langer, 2002) on reflective practices clearly shows that many adults do not necessarily know how to reflect.

The important work of Argyris and Schön (1996) on introducing and sustaining individual learning, specifically using double-loop learning, should be used when implementing an organizational learning program. Ravell also showed us that time is an important factor for individual development and that political factions are part of that process. With patience and an ongoing program, group learning activities can be introduced to operations personnel, thereby supporting the kind of system-level thinking proposed by Senge (1990).

A critical part of organizational learning, in particular the necessary steps to establish a learning organization, is the formation of communities of practice. Communities of practice, in all of the case studies, were the cornerstones in the transition from individual-based learning to group learning. Communities of practice begin the maturation process of getting organizations to change to learning based on organizational goals. This is critical for ROD because technology requires planning and vision that are consistent with business strategy. While much of the literature integrates the notion of communities of practice with knowledge management, I expanded its use and defined the community as the single most important organizational structure for dealing with emerging technologies. The reason for this is the very challenge facing IT organizations today: to be able to integrate their efforts across business units. This has been proven to be the most difficult challenge for the chief information officers (CIOs) of today. This was further supported by the Siemens AG case study, in which Dana Deasy, the corporate CIO of the Americas, provided a detailed picture of the complex world of a CIO in a global firm, with over 400,000 employees. Yet, it is the creation of multiple layers of communities of practice that enables firms to create what I call "common threads" of communication. Thus, the linkage across communities of practice is a central theme of this book, providing guidance and education to organizations to establish processes that support their evolution in a responsive way.

The key word that I have used here is *evolution*. In the past, information traveled much slower, and there was more time to interpret its impact on the organization. Today, the travel time is shrinking; therefore, evolution can and should occur at a quicker pace. Indeed, organizational evolution is intertwined with the dynamics of community legitimization (Aldrich, 2001). Technological development for a particular population has widespread consequences for the rest of the organization. In these cases, technological innovations represent a form of collective learning that is different from direct learning from experience alone (Miner & Haunschild, 1995). There are many scholars who believe that change management must be implemented through top-down management approaches. However, I hope this book demonstrates that leadership through top-down management will never be solely sufficient to establish the organizational structure needed to handle technological innovations properly. Many such efforts to reorganize or reengineer organizations have had disappointing results. Many of these failures, I believe, are attributable to a dependence on management intervention as opposed to strategic integration and cultural assimilation. Technology only serves to expose problems that have existed in organizations for decades: the inability to drive down responsibilities to the operational levels of the organization.

My case studies provide, I trust, a realistic and pragmatic view toward the attainment of responsive organizational dynamism, assuming the appropriate roles and responsibilities are available. Furthermore, the case studies also reflect that progress toward organizational learning and maturity is a gradual one. As such, I determined that organizational transformation must be addressed along the same basis; that is, transformation is a gradual process as opposed to a planned specific outcome. I showed that organizations could and should look at transformation in much shorter "chunks," as opposed to long-term "big-bang" approaches that rarely work and are difficult to measure. Measurement was applied to organizational transformation via the implementation of the balanced scorecard. The scorecard model I modified is tied to the chunk approach.

Another important concept in this book is the reconciliation between control and empowerment. As organizations find that their traditional structures are cumbersome when dealing with emerging technologies, they realize the need to empower employees to do more dynamically. With this empowerment, employees may make more mistakes or seem less genuine at times. When this occurs, there may be a need for management controls to be instituted and power-centralized management styles to be incorporated. Unfortunately, too many controls end any hope of creating a learning organization that can foster the dynamic planning and needs of responsive organizational dynamism. They also block the molding of communities of practice that require common threads of discourse and language. Indeed, it is communities of practice and discourse that lay the foundations for addressing the dilemma of employee control versus empowerment.

We are really beginning to experience the results of emerging technologies, particularly for products traded internationally. We have seen an unusual trend occur in which off shore product development and maintenance is at an all-time high, local employment is down, and corporate earnings are growing. The advent of this cycle lays a foundation for the new trends of global worker operations, many of which are shifting from a labor-intensive process to needs for thinking, planning, and management.

Unskilled or less-skilled workers, partly because of new technological automation, are allowing organizations to displace higher-costing local labor to international outsourced operations. This means an increase in management learning related to supervision and coordination in a technology-driven world. We must be aware of the concern expressed by O'Sullivan (2001) that "new technologies have created unemployed workers with no rights" (p. 159). The way individuals communicate, or the rules of their engagement, are quickly changing, particularly in the need to create more research and development (R&D) infrastructures that can respond quicker to innovation opportunities brought about by emerging technologies. We saw this dilemma occur at Siemens, where business strategy and technology became a major investment, and the realization that e-business was more about business than just technology.

To address the lack of understanding of the technology life cycle, I presented my concept of driver and supporter functions and mapped them onto evolutionary transformation. This life cycle is one that ties business strategy into technology and should be used to convey ROD to executives. Driver functions explain why strategic integration is so important and present a case that requires more marketing-based philosophies when investing in technologies. This means that early adaptation of technology requires, as Bradley and Nolan (1998) call it, "sense-and-respond" approaches, by which IT organizations can experiment with business units on how a technology may benefit the business. Siemens and ICAP provided good examples of different ways of creating infrastructures that can support technology exploration, including Deasy's 90-day program, by which technology investments were reviewed periodically to see what adjustments are required to maximize the investment. It also provided a way to cancel those investments that were not paying off as originally forecast. Understanding that changes along the way are needed, or that there are technologies that do not provide the intended benefits, must become a formal part of the process, one that CEOs must recognize and fund.

On the other hand, the supporter role is one that addresses the operational side of IT, such that executives and managers understand the difference. I treated the concept of supporter as an eventual reality, the reality that all technologies, once adopted by operations, must inevitably become a commodity. This realization paves the way to understanding when and how technologies can be considered for outsourcing, based on economies of scale. The adoption of this philosophy creates a structured way of understanding the cost side of the IT dilemma and requires business units to integrate their own plans with those offered by emerging technologies. The supporter aspect of technology became the base of cultural assimilation because once a technology is adapted by operations, there must be a corresponding process that fosters its impact on organizational structures and cultural behaviors. It also provides the short- and long-term expected transformations, which ultimately link technology and strategic performance.

The driver/supporter philosophy also shows the complexity of the many definitions of technology, and that executives should not attempt to oversimplify it. Simply put, technology must be discussed in different ways, and chief IT executives need to rise to the occasion to take a leadership role in conveying this to executives, managers, and operations, through organizational learning techniques. Organizations that can implement driver/supporter methods will inevitably be better positioned to understand why they need to invest in certain technologies and technology projects. My initial case study at Ravell exposed the potential limit of only operating on the unit levels and not getting executives involved in the system thinking and learning phases.

These general themes can be formulated as a marriage between business strategy and technological innovation and can be represented as follows:

1. Organizations must change the business cycles of technology investment; technology investment must become part of the everyday or normative processes, as opposed to specific cycles based on economic opportunities or shortfalls. Emerging technologies tend to be implemented on a "stop-and-go" basis, or based on breakthroughs, followed by discontinuities (Tushman & Anderson, 1997).

2. The previous experiences that organizations have had with technology are not a good indicator for its future use. Technology innovations must evolve through infrastructure, learning, and process evaluation.

3. Technology is central to competitive strategy. Executives need to ensure that technology opportunities are integrated with all discussions on business strategy.

4. Research and development (R&D) is at the center of systems/organizational-level thinking and learning. Companies need to create R&D operations, not as separate entities, but as part of the evaluation processes within the organizational structure.

5. Managing technology innovations must be accomplished through linkages. Thus, interfaces across communities of practice via common threads are essential to have learning improve the ability of the organization to operate within responsive organizational dynamism.

6. Managing intellectual capital is an exercise of linking the various networks of knowledge in the organization. Managing this knowledge requires organizational learning, to transfer tacit knowledge to explicit knowledge. The cultural assimilation component of ROD creates complex tacit knowledge between IT and non-IT business units.

7. There are multiple and complex levels of management that need to be involved in responsive organizational dynamism. Successful management utilizes organizational learning practices to develop architectures, manage change, and deal with

short- and long-term projects simultaneously. Strong leadership will understand that the communities of practice among the three primary levels (executive, middle management, and operations) constitute the infrastructure that best sustains the natural migration toward responsive organizational dynamism.

This book looked at business strategy from yet another perspective, beyond its relationship with emerging technologies. Because organizational learning is required to foster responsive organizational dynamism, strategy must also be linked to learning. This linkage is known as *strategic learning,* which, if implemented, helps organizations to continually adapt to the changing business environment, including changes brought about by technology.

However, due to the radical speed, complexity, and uncertainty, traditional ways of doing strategy and learning can no longer ignore the importance of technology. The old methods of determining business strategy were based on standard models that were linear and "plug-in." As stated, they were also very much based on projects that attempted to design one-time efforts with a corresponding result. As Pietersen (2002) explains, "These processes usually produce operating plans and budgets, rather than insights and strategic breakthroughs" (p. 250). Technological dynamism has accelerated the need to replace these old traditions, and I emphasized that organizations that practice ROD must

- Evaluate and implement technology in an ongoing process and embed it as part of normal practices. This requires a change in integration and culture.
- Comprehend that the process of ROD is not about planning; it is about adaptation and strategic innovation.
- Have a process that feeds on the creation of new knowledge through organizational learning toward strategic organizational transformation.

Many scholars might correlate strategic success with leadership. While leadership, in itself, is an invaluable variable, it is just that. To attain ongoing evolution, I believe we need to move away from relying on individual leadership efforts and move toward an infrastructure that has fewer leaders and more normative behavior that can support and sustain responsive organizational dynamism. Certainly, this fosters the important roles and responsibilities of CEOs, managers, and boards, but to have an ongoing process that changes the thinking and the operational fundamentals of the way the organization functions is more important and more valuable than individual leadership. That is why I raised the issues of discourse and language as well as self-development. Therefore, it is the ability of an organization to transform its entire community that will bring forth long-term strategic performance.

What this book really commits to is the importance of lifelong learning. The simple concept is that adults need to continually challenge their cultural norms if they are to develop what Mezirow (1990) calls "new meaning perspectives." It is these new meaning perspectives that lay the foundation for ROD so that managers and staff can continually challenge themselves to determine if they are making the best strategic decisions. Furthermore, it prepares individuals to deal with uncertainty as well as the ongoing transitions in the way they do their jobs. It is this very process that ultimately fosters learning in organizations.

While on-the-job training is valuable, Ravell shows us that movement, or rotation of personnel, often supports individual learning. Specifically, the relocation of IT personnel to a business unit environment during Ravell phase I served to get IT staff more acclimated to business issues. This relocation helped IT staff members to begin to reflect about their own functions and their relationship to the overall mission of the organization. Ravell phase III showed yet another transition; taking a group of IT staff members and permanently integrating them in a non-IT business-specific department. Ravell also teaches us that reflection must be practiced; time must be devoted to its instruction, and it will not occur automatically without interventions from the executive rank. The executive must be a "champion" who demonstrates to staff that the process is important and valued. Special sessions also need to be scheduled that make the process of learning and reflection more formal. If this is done and nurtured properly, it will allow communities to become serious about best practices and new knowledge creation.

Although I used technology as the basis for the need for responsive organizational dynamism, the needs for its existence can be attributed to any variable that requires dynamic change. As such, I suggest that readers begin to think about the next "technology" or variable that can cause the same needs to occur inside organizations. Such accelerations are not necessarily limited to technology. For example, we are experiencing the continuation of organizational downsizing from acquisitions. These acquisitions present similar challenges in that organizations must be able to integrate new cultures and "other" business strategies and attempt to form new holistic directions—directions that need to be formed quickly to survive.

The market per se also behaves in a similar way to technology. The ability to adjust to consumer needs and shifting market segments is certainly not always related to technological change. My point is that ROD is a concept that should be embraced notwithstanding whether technology seems to have slowed or to have no effect on a specific industry at a particular moment. Thus, I challenge the organizations of today to develop new strategies that embrace the need to become dynamic throughout all of their operations and to create communities of practice that plan for ongoing strategic integration and cultural assimilation.

This book looked at the advent of technology to uncover a dilemma that has existed for some time. Perhaps a more general way of defining what ROD offers is to compare it to another historical concept: "self-generating organizations." Self-generating organizations are known for their promotion of autonomy with an "underlying organic sense of interdependence" (Johansen et al., 1995). Based on this definition, a self-generating organization is like an organism that evolves over time. This notion is consistent with organizational learning because they both inherently support inner growth stemming from the organization as opposed to its executives. The self-generating organization works well with ROD in the following ways:

- Traditional management control systems do not apply.
- Risks are higher, given that these organizational workers are granted a high degree of autonomy and empowerment that will lead to processes that break with the norms of the business.

- Adjustments and new processes should be expected.
- These organizations tend to transform political activity into strong supporting networks.
- Leadership definitions do not work. You cannot lead what you cannot control.

Self-generating organizations have scared traditional managers in the past, due to the fear they have of losing control. ROD provides a hybrid model that allows for self-generating infrastructures while providing certain levels of control fostered by organizational learning. Specifically, this means that the control is not traditional control. Responsive organizational dynamism, for example, embraces the breaking of rules for good reasons; it allows individuals to fail yet to reflect on the shortfall so that they do not repeat the same errors. It also allows employees to take risks that show promise and lead to increased critical thinking and to strategic action. Indeed, management and leadership become more about framing conditions for operations, observing the results, and making adjustments that maintain stability. Thus, seeing ROD as a form of self-generation is the basis for sustaining innovative infrastructures that can respond to dynamic variables, like technology.

I have emphasized the need for organizational learning as the key variable to make ROD a reality. While I have modified many of the organizational learning theories to fit this need, I must acknowledge that a portion of the "learning" should be considered "organizing." Vince (2002) provides an analysis of how organizational learning could be used to sustain an "organized" reflection. He provides an interesting matrix of how the two theories can be integrated. After reviewing many of the ways in which organizational learning affects responsive organizational dynamism, I have developed a modified chart of Vince's original framework, as shown in Table 13.1.

Table 13.1 shows the three kinds of reflective practices that can operate in an organization: individual, group, and organizational. I emphasized in Chapter 9 that the extent of organizational learning maturation is directly related to the sophistication of reflections among the communities of practice. The more learning that occurs, based on individual reflection, the earlier the stage at which organizational learning maturity occurs. Thus, more mature organizations

Table 13.1 Individual, Group, and Organizational and Reflective Practices

	INDIVIDUAL RELATIONS BETWEEN THE PERSON, ROLE, AND THE ORGANIZATION-IN-THE-MIND	GROUP RELATIONS ACROSS THE IT BOUNDARIES OF SELF OR OTHER AND OF SUBDEPARTMENTS WITHIN IT	ORGANIZATIONAL THE RELATIONS BETWEEN IT AND OTHER BUSINESS UNITS
Peer consulting groups (nonmanagerial self-governing IT groups of at least three individuals)	Making connections for the self: Review and reflection within IT community, by friendship, and mutuality of interests and needs.	Making connections in small groups with "others" across IT organization: Develop interpersonal communication and dialogue within IT communities.	Making connections with the entire organization: Reflection on ways that technology affects other groups in the organization.
Organizational role analysis (linking individuals with "others" inside the IT organization)	Organizational role analysis: Understanding the connections between the person, the person in IT, and his or her role in organization.	Role analysis groups: The ways in which technology roles and the understanding of those roles interweave within an IT community or department.	Technology role provides the framework within which the person and organization are integrated.
Communities of practice (groups of individuals united in actions that contribute to the production of IT ideas in practice)	Involvement: Providing personal experience as the vehicle to organize the use of technology.	Engagement: Experience used to apply technology across IT organization; understanding of importance of IT interdepartment communication.	Establishment: Experience of power relations as they react and respond to technology uses among communities of practice
Group relations conferences (reveal the complexities of feelings, interactions, and power relations that are integral to the process of organizing technology implementations)	Experiencing and rethinking technology authority and the meaning and consequences of leadership and followership.	Experiencing defensive mechanisms and avoidance strategies across IT departments. Experience of organizing, belonging, and representing across IT organizations.	Experiencing the ways in which IT and the organization become integrated using collective emotional experience, politics, leadership, authority, and organizational transformation.

reflect at the group and organizational level. Becoming more mature requires a structured process that creates and maintains links between reflection and democratic thinking. These can be mapped onto the ROD arc, showing how, from an "organizing" perspective, reflective practice serves as a process to "outline what is involved in the process of reflection for learning and change" (Vince, 2002, p. 74). Vince does not, however, establish a structure for implementation, for which ROD serves that very purpose, as shown in Figure 13.2.

Figure 13.2 graphically shows how organized reflection maps to the linear stages of the ROD arc (the organizational-level maturity arc), which in turn maps onto the three best practices arcs, discussed in Chapter 9. Each of the management arcs represents a level of management maturity at the organizational level, with Vince's (2002) matrix providing the overarching concepts on how to actually organize the progression from individual-based thinking and reflection to a more comprehensive and systems-level thinking and learning base.

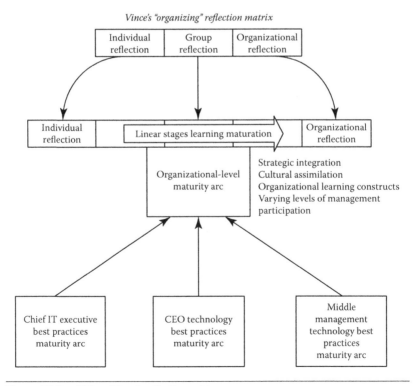

Figure 13.2   ROD and Vince's reflection matrix.

The emphasis, overall, is that individual learning alone will undermine collective governance. Therefore, the movement from individual to organizational self-management remains a critical part of understanding how technology and other dynamic variables can foster new strategies for competitive advantage.

Perhaps the most important conclusion of this third edition is the impact that digital technologies are having on the acceleration of change being experienced throughout the world. Indeed, digital technology has begun to change not only the business world but the very fabric of our lives. Particular to this change is the continual emergence of social media as a driver of new and competitive products and services. I also discussed the changing work philosophy and expectations of our new generation of employees, and how they think differently and want a more complex experience in the places in which they work. The Gen Y population is clearly a new breed of employees and the Gen Z behind them will be even more accustomed to using digital technologies in every fabric of the ways they want to learn, their preferences in communicating with others, and their role in society. Most important are the ways that technology has changed consumer behavior. I truly believe that future generations will look back on this period and indeed say, this was truly a consumer revolution!

Glossary

baby boomers: The generation of individuals who were born between the years of 1946 and 1964.

business process reengineering: a process that organizations undertake to determine how best to use technology to improve business performance

customer relationship management (CRM): the development and maintenance of integrated relationships with the customer base of an organization. CRM applications provide organizations with integrated tools that allow individuals to store and sustain valuable information about their customers.

data mapping: the process of comparing the data fields in one database to another, or toward a new application database

decision-support systems (DSS): systems that assist managers to make better decisions by providing analytical results from stored data

digital disruption: When new digital technology advancements impact the value of goods and services.

digital transformation: The repositioning of or a new investment in technology and business models in efforts to compete in a rapidly changing digital economy and create a newfound sense of value for customers.

enterprise resource planning (ERP): a set of multimodule applications that support an entire manufacturing and business operation, including product planning, purchasing, maintaining inventories, interacting with suppliers, providing customer service, accounting interfaces, and tracking order shipments. These systems are also known as enterprise-level applications.

garbage can: an abstract concept for allowing individuals a place to suggest innovations, brought about by technology. The inventory of technology opportunities needs regular evaluation

Gen X: The generation of individuals who were born between the years of 1965 and 1980.

Gen Y/Millennials: The generation of individuals who were born between the years of 1981 and 1992. There is disagreement on the exact end dates of Gen Y individuals.

internet: a cooperative message-forwarding system that links computer networks all over the world.

intranet: a network confined to a single organization or unit.

ISO 9000: a set of quality assurance standards published by the 91-nation International Organization for Standardization (ISO). ISO 9000 requires firms to define and implement quality processes in their organization.

legacy: an existing software application or system that is assumed to operate. By definition, all applications in production become legacies.

operational excellence: a philosophy of continuous improvement throughout an organization by enhancing efficiency and quality across operations

outsourcing: A practice utilized by corporations which involves having external suppliers complete internal work in efforts to reduce costs.

storyboarding: the process of creating prototypes that allow users to actually see examples of technology, and how it will look and operate. Storyboarding tells a story and can quickly educate executives, without being intimidating.

technology definitions branding: the process of determining how an organization wants to be viewed by its customers. Branding includes not only the visual view, but also the emotional, rational, and cultural image that consumers associate with an organization, its products, and services.

user interface: the relationship with end users that facilitates the process of gathering and defining logical requirements

user level: the tier of computer project experience of the user. There are three levels: (1) knowledgeable, (2) amateur, and (3) novice

virtual teams: groups of people, geographically disbursed, and linked together using communication technologies

World Wide Web (web): loosely organized set of computer sites that publish information that anyone can read via the Internet using mainly HTTP (Hypertext Transfer Protocol)

Year 2000 (Y2K): a monumental challenge to many organizations due to a fear that software applications could not handle the turn of the century. Specifically, calculations that used the year portion of a date would not calculate properly. As such, there was a huge investment in reviewing legacy systems to uncover where these flaws existed.

Organizational Learning Definitions

action science: pioneered by Argyris and Schön (1996), action science was designed to promote individual self-reflection, regarding behavior patterns and to encourage a productive exchange among individuals. Action science encompasses a range of methods to help individuals learn how to be reflective about their actions. A key component of action science is the use of reflective practices—including what is commonly known among researchers and practitioners as reflection in action, and reflection on action.

balanced scorecard: a means for evaluating transformation, not only for measuring completion against set targets, but also, for defining how expected transformations map onto the strategic objectives of the organization. In effect, it is the ability of the organization to execute its strategy.

communities of practice: are based on the assumption that learning starts with engagement in social practice and that this practice is the fundamental construct by which individuals learn.

Thus, communities of practice are formed to get things done using a shared way of pursuing interest.

cultural assimilation: a process that focuses on the organizational aspects of how technology is internally organized, including the role of the IT department, and how it is assimilated within the organization as a whole. It is an outcome of responsive organizational dynamism.

cultural lock-in: the inability of an organization to change its corporate culture, even when there are clear market threats (Foster & Kaplan, 2001)

double-loop learning: requires individuals to reflect on a prior action or habit that needs to change in behavior and change to operational procedures. For example, people who engage in double-loop learning, may need to adjust how they perform their job as opposed to just the way they communicate with others.

drivers: those units that engage in frontline or direct revenue-generating activities

experiential learning: a type of learning that comes from the experiences that adults have accrued over the course of their individual lives. These experiences provide rich and valuable forms of "literacy" that must be recognized as important components to overall learning development.

explicit knowledge: documented knowledge found in manuals, documentation, files, and other accessible places and sources

flame: a lengthy, often personally insulting, debate in an electronic community that provides both positive and negative consequences

frame-talk: focuses on interpretation to evaluate the meanings of talk

knowledge management: the ability to transfer individual tacit knowledge into explicit knowledge

left-hand column: a technique by which individuals use the right-hand column of a piece of paper to transcribe dialogues that they feel have not resulted in effective communication. In the left-hand column of the same page, participants write what they were really thinking at the time of the dialogue but did not say.

management self-development: increases the ability and willingness of managers to take responsibility for themselves, particularly for their own learning (Pedler et al., 1988)

mythopoetic-talk: communicates ideogenic ideas and images that can be used to communicate the nature of how to apply tool-talk and frame-talk, within the particular culture or society. This type of talk allows for concepts of intuition and ideas for concrete application.

organizational knowledge: is defined as "the capability of a company as a whole to create new knowledge, disseminate it throughout the organization, and embody it in products, services, and systems" (Nonaka & Takeuchi, 1995, p. 3)

organizational transformation: changes in goals, boundaries, and activities. According to Aldrich (2001), organizational transformations "must involve a qualitative break with routines and a shift to new kinds of competencies that challenge existing organizational knowledge" (p. 163).

reflection with action: term used as a rubric for the various methods involving reflection in relation to activity

responsive organizational dynamism: the set of integrative responses, by an organization, to the challenges raised by technology dynamism. It has two component outcomes: strategic integration and cultural assimilation.

single-loop learning: requires individuals to reflect on a prior action or habit that needs to be changed in the future but that does not require individuals to change their operational procedures with regard to values and norms

strategic integration: a process that addresses the business-strategic impact of technology on organizational processes. That is, the business-strategic impact of technology requires immediate organizational responses and, in some instances, zero latency. It is an outcome of responsive organizational dynamism, and it requires organizations to deal with a variable that forces acceleration of decisions in an unpredictable fashion.

supporters: units that do not generate obvious direct revenues but rather are designed to support frontline activities

tacit knowledge: an experience-based type of knowledge and skill, with the individual capacity to give intuitive forms to new things; that is, to anticipate and preconceptualize the future (Kulkki & Kosonen, 2001)

technological dynamism: characterizes the unpredictable and accelerated ways in which technology, specifically, can change strategic planning and organizational behavior/culture. This change is based on the acceleration of events and interactions within organizations, which in turn create the need to better empower individuals and departments.

tool-talk: includes instrumental communities required to discuss, conclude, act, and evaluate outcomes

References

Aldrich, H. (2001). *Organizations Evolving*. London: Sage.

Allen, F., & Percival, J. (2000). Financial strategies and venture capital. In G. S. Day & P. J. Schoemaker (Eds.), *Wharton on Managing Emerging Technologies* (pp. 289–306). New York: Wiley.

Allen, T. J., & Morton, M. S. (1994). *Information Technology and the Corporation*. New York: Oxford University Press.

Applegate, L. M., Austin, R. D., & McFarlan, F. W. (2003). *Corporate Information Strategy and Management* (2nd edn.). New York: McGraw-Hill.

Argyris, C. (1993). *Knowledge for Action: A Guide to Overcoming Barriers to Organizational Change*. San Francisco, CA: Jossey-Bass.

Argyris, C., & Schön, D. A. (1996). *Organizational Learning II*. Reading, MA: Addison-Wesley.

Arnett, R. C. (1992). *Dialogue Education: Conversation about Ideas and between Persons*. Carbondale, IL: Southern Illinois University Press.

Bakanauskiené, I., Bendaravicliené, R., Krikstolaitis, R., & Lydeka, Z. (2011). Discovering an employer branding: Identifying dimensions of employers' attractiveness in University. *Management of Organizations: Systematic Research*, 59, pp. 7–22.

Bazarova, N. N., & Walther, J. B. (2009). Virtual groups: (Mis)attribution of blame in distributed work. In P. Lutgen-Sandvik & B. Davenport Sypher (Eds.), *Destructive Organizational Communication: Processes, Consequences, and Constructive Ways of Organizing* (pp. 252–266). New York: Routledge.

Bellovin. S. M. (2015). *Thinking Security: Stopping Next Year's Hackers*. Boston, MA: Addison-Wesley.

Bensaou, M., & Earl, M. J. (1998). The right mind-set for managing information technology. In J. E. Garten (Ed.), *World View: Global Strategies for the New Economy* (pp. 109–125). Cambridge, MA: Harvard University Press.

Bertels, T., & Savage, C. M. (1998). Tough questions on knowledge management. In G. V. Krogh, J. Roos, & D. Kleine (Eds.), *Knowing in Firms: Understanding, Managing and Measuring Knowledge* (pp. 7–25). London: Sage.

Boland, R. J., Tenkasi, R. V., & Te'eni, D. (1994). Designing information technology to support distributed cognition. *Organization Science, 5,* 456–475.

Bolman, L. G., & Deal, T. E. (1997). *Reframing Organizations: Artistry, Choice, and Leadership* (2nd edn.). San Francisco, CA: Jossey-Bass.

Bradley, S. P., & Nolan, R. L. (1998). *Sense and Respond: Capturing Value in the Network Era.* Boston, MA: Harvard Business School Press.

Brown, J. S., & Duguid, P. (1991). Organizational learning and communities of practice. *Organization Science, 2,* 40–57.

Burke, W. W. (2002). *Organizational Change: Theory and Practice.* London: Sage.

Cadle, J., Paul, D., & Turner, P. (2014). *Business Analysis Techniques: 99 Essential Tools for Success* (2nd edn.). Swindon, UK: Chartered Institute for IT.

Capgemini Consulting. (2013). *Being Digital: Engaging the Organization to Accelerate Digital Transformation* [White Paper]. Retrieved from http://www.capgemini-consulting.com/resource-file-access/resource/pdf/being_digital_engaging_the_organization_to_accelerate_digital_transformation.pdf

Cash, J. I., & Pearlson, K. E. (2004, October 18). The future CIO. *Information Week.* Available at http://www.informationweek.com/story/showArticle.jhtml?articleID=49901186

Cassidy, A. (1998). *A Practical Guide to Information Strategic Planning.* Boca Raton, FL: St. Lucie Press.

Cisco (2012). *Creating an Office from an Easy Chair* [White Paper]. Retrieved from http://www.cisco.com/c/en/us/solutions/collateral/enterprise/cisco-on-cisco/Trends_in_IT_Gen_Y_Flexible_Collaborative_Workspace.pdf

Collis, D. J. (1994). Research note: How valuable are organizational capabilities? *Strategic Management Journal, 15,* 143–152.

Cross, T, & Thomas, R. J. (2009). *Driving Results through Social Networks. How Top Organizations Leverage Networks for Performance and Growth.* San Francisco, CA: Jossey-Bass.

Cyert, R. M., & March, J. G. (1963). *The Behavioral Theory of the Firm.* Englewood Cliffs, NJ: Prentice-Hall.

De Jong, G. 2014. Financial inclusion of youth in the Southern provinces of Santander: Setting up a participatory research in Columbia. In P. Wabike & J. van der Linden (Eds.), *Education for Social Inclusion* (pp. 87–106). Groningen, the Netherlands: University of Groningen.

Dewey, J. (1933). *How We Think.* Boston, MA: Health.

Dodgson, M. (1993). Organizational learning: A review of some literatures. *Organizational Studies, 14,* 375–394.

Dolado, J. (Ed.) (2015). *No Country for Young People? Youth Labour Market Problems in Europe,* VoxEU.org eBook, London: CEPR Press.

Dragoon, A. (2002). This changes everything. Retrieved December 15, 2003, from http://www.darwinmag.com

Earl, M. J. (1996a). Business processing engineering: A phenomenon of organizational dimension. In M. J. Earl (Ed.), *Information Management: The Organizational Dimension* (pp. 53–76). New York: Oxford University Press.

Earl, M. J. (1996b). *Information Management: The Organizational Dimension.* New York: Oxford University Press.

Earl, M. J., Sampler, J. L., & Short, J. E. (1995). Strategies for business process reengineering: Evidence from field studies. *Journal of Management Information Systems, 12,* 31–56.

Easterby-Smith, M., Araujo, L., & Burgoyne, J. (1999). *Organizational Learning and the Learning Organization: Developments in Theory and Practice.* London: Sage.

Eisenhardt, K. M., & Bourgeois, L. J. (1988). Politics of strategic decision making in high-velocity environments: Toward a midrange theory. *Academy of Management Journal, 31,* 737–770.

Elkjaer, B. (1999). In search of a social learning theory. In M. Easterby-Smith, J. Burgoyne, & L. Araujo (Eds.), *Organizational Learning and the Learning Organization* (pp. 75–91). London: Sage.

Ernst & Young, (2012). The digitization of everything: How organisations must adapt to changing consumer behaviour [White Paper]. Retrieved from http://www.ey.com/Publication/ vwLUAssets/The_digitisation_ of_everything_-_How_organisations_must_adapt_to_changing_consumer_behaviour/$FILE/ EY_Digitisation_of_everything.pdf

Fineman, S. (1996). Emotion and subtexts in corporate greening. *Organization Studies, 17,* 479–500.

Foster, R. N., & Kaplan, S. (2001). *Creative Destruction: Why Companies That Are Built to Last Underperform the Market: And How to Successfully Transform Them.* New York: Currency.

Franco, V., Hu, H., Lewenstein, B. V., Piirto, R., Underwood, R., & Vidal, N. K. (2000). Anatomy of a flame: Conflict and community building on the Internet. In E. L. Lesser, M. A. Fontaine, & J. A. Slusher (Eds.), *Knowledge and Communities* (pp. 209–224). Woburn, MA: Butterworth-Heinemann.

Friedman, T. L. (2007). *The World Is Flat.* New York: Picador/Farrar, Straus and Giroux.

Friedman, V. J, Razer, M., Tsafrir, H., & Zorda, O. (2014). An action science approach to creating inclusive teacher-parent relationships. In P. Wabike & J. van der Linden (eds.), *Education for Social Inclusion* (pp. 25–51). Groningen, the Netherlands: University of Groningen.

Garvin, D. A. (1993). Building a learning organization. *Harvard Business Review*, *71*(4), 78–84.

Garvin, D. A. (2000). *Learning in Action: A Guide to Putting the Learning Organization to Work.* Boston, MA: Harvard Business School Press.

Gephardt, M. A., & Marsick, V. J. (2003). Introduction to special issue on action research: Building the capacity for learning and change. *Human Resource Planning*, *26*(2), 14–18.

Glasmeier, A. (1997). The Japanese small business sector (Final report to the Tissot Economic Foundation, Le Locle, Switzerland, Working Paper 16). Austin: Graduate Program of Community and Regional Planning, University of Texas at Austin.

Grant, D., Keenoy, T., & Oswick, C. (Eds.). (1998). *Discourse and Organization.* London: Sage.

Grant, R. M. (1996). Prospering in a dynamically-competitive environment: Organizational capability as knowledge integration. *Organization Science*, *7*, 375–387.

Gregoire, J. (2002, March 1). The state of the CIO 2002: The CIO title, What's it really mean? *CIO.* Available at http://www.cio.com/article/30904/The_State_of_the_CIO_2002_The_CIO_Title_What_s_It_Really_Mean_

Habermas, J. (1998). *The Inclusion of the Other: Studies in Political Theory.* Cambridge, MA: MIT Press.

Halifax, J. (1999). Learning as initiation: Not-knowing, bearing witness, and healing. In S. Glazier (Ed.), *The Heart of Learning: Spirituality in Education* (pp. 173–181). New York: Penguin Putnam.

Hardy, C., Lawrence, T. B., & Philips, N. (1998). Talk and action: Conversations and narrative in interorganizational collaboration. In D. Grant, T. Keenoy, & C. Oswick (Eds.), *Discourse and Organization* (pp. 65–83). London: Sage.

Heath, D. H. (1968). *Growing Up in College: Liberal Education and Maturity.* San Francisco, CA: Jossey-Bass.

Hoffman, A. (2008, May 19). The social media gender gap. *Business Week.* Available at http://www.businessweek.com/technology/content/may2008/tc20080516_580743.htm

Huber, G. P. (1991). Organizational learning: The contributing processes and the literature. *Organization Science*, *2*, 99–115.

Hullfish, H. G., & Smith, P. G. (1978). *Reflective Thinking: The Method of Education.* Westport, CT: Greenwood Press.

Huysman, M. (1999). Balancing biases: A critical review of the literature on organizational learning. In M. Easterby-Smith, J. Burgoyne, & L. Araujo (Eds.), *Organizational Learning and the Learning Organization* (pp. 59–74). London: Sage.

Johansen, R., Saveri, A., & Schmid, G. (1995). Forces for organizational change: 21st century organizations: Reconciling control and empowerment. *Institute for the Future*, *6*(1), 1–9.

Johnson Controls (2010) Generation Y and the workplace: Annual report 2010 [White Paper]. Retrieved from http://www.johnsoncontrols.com/content/dam/WWW/jci/be/global_workplace_innovation/oxygenz/Oxygenz_Report_-_2010.pdf

Jonas, L., & Kortenius, R. (2014) Beyond a Paycheck: Employment as an act of consumption for Gen Y talents (Master's Thesis) Retrieved from http://lup.lub.lu.se/luur/download?func=downloadFile&recordOId=4456566&fileOId=4456569

Jones, M. (1975). Organizational learning: Collective mind and cognitivist metaphor? *Accounting Management and Information Technology*, *5*(1), 61–77.

Kanevsky, V., & Housel, T. (1998). The learning-knowledge-value cycle. In G. V. Krogh, J. Roos, & D. Kleine (Eds.), *Knowing in Firms: Understanding, Managing and Measuring Knowledge* (pp. 240–252). London: Sage.

Kaplan, R. S., & Norton, D. P. (2001). *The Strategy-Focused Organization*. Cambridge, MA: Harvard University Press.

Kegan, R. (1994). *In Over Our Heads: The Mental Demands of Modern life*. Cambridge, MA: Harvard University Press.

Kegan, R. (1998, October). *Adult Development and Transformative Learning*. Lecture presented at the Workplace Learning Institute, Teachers College, New York.

Knefelkamp, L. L. (1999). Introduction. In W. G. Perry (Ed.), *Forms of Ethical and Intellectual Development in the College Years: A Scheme* (pp. xi–xxxvii). San Francisco, CA: Jossey-Bass.

Koch, C. (1999, February 15). Staying alive. *CIO Magazine*. 38–45.

Kolb, D. (1984a) *Experiential Learning: Experience as the Source of Learning and Development*. Englewood Cliffs, NJ: Prentice-Hall.

Kolb, D. (1984b). *Experiential Learning as the Science of Learning and Development*. Englewood Cliffs, NJ: Prentice Hall.

Kolb, D. (1999). *The Kolb Learning Style Inventory*. Boston, MA: HayResources Direct.

Kulkki, S., & Kosonen, M. (2001). How tacit knowledge explains organizational renewal and growth: The case at Nokia. In I. Nonaka & D. Teece (Eds.), *Managing Industrial Knowledge: Creation, Transfer and Utilization* (pp. 244–269). London: Sage.

Langer, A. M. (1997). *The Art of Analysis*. New York: Springer-Verlag.

Langer, A. M. (2001a). *Analysis and Design of Information Systems*. New York: Springer-Verlag.

Langer, A. M. (2001b). Fixing bad habits: Integrating technology personnel in the workplace using reflective practice. *Reflective Practice*, *2*(1), 100–111.

Langer, A. M. (2002). Reflecting on practice: using learning journals in higher and continuing education. *Teaching in Higher Education*, *7*, 337–351.

Langer, A. M. (2003). Forms of workplace literacy using reflection-with action methods: A scheme for inner-city adults. *Reflective Practice*, *4*, 317–336.

Langer, A. M. (2007). *Analysis and Design of Information Systems* (3rd edn.). New York: Springer-Verlag.

Langer A. M. (2009). Measuring self-esteem through reflective writing: Essential factors in workforce development. *Journal of Reflective Practice*, *9*(10): 45–48.

Langer, A. (2011). *Information Technology and Organizational Learning: Managing Behavioral Change through Technology and Education* (2nd edn.). Boca Raton, FL: CRC Press.

Langer, A. M. (2013). Employing young talent from underserved populations: Designing a flexible organizational process for assimilation and productivity. *Journal of Organization Design*, *2*(1): 11–26.

Langer, A. M. (2016) *Guide to Software Development: Designing and Managing the Life Cycle* (2nd edn.). New York: Springer.

Langer, A.M. & Yorks, L. 2013. *Strategic IT: Best practices for Managers and Executives*. Hoboken, NJ: Wiley.

Lesser, E. L., Fontaine, M. A., & Slusher, J. A. (Eds.). (2000). *Knowledge and Communities*. Woburn, MA: Butterworth-Heinemann.

Levine, R., Locke, C., Searls, D., & Weinberger, D. (2000). *The Cluetrain Manifesto*. Cambridge, MA: Perseus Books.

Lientz, B. P., & Rea, K. P. (2004). *Breakthrough IT Change Management: How to Get Enduring Change Results*. Burlington, MA: Elsevier Butterworth-Heinemann.

Lipman-Blumen, J. (1996). *The Connective Edge: Leading in an Independent World*. San Francisco, CA: Jossey-Bass.

Lipnack, J., & Stamps, J. (2000). *Virtual Teams* (2nd edn.). New York: Wiley.

Lounamaa, P. H., & March, J. G. (1987). Adaptive coordination of a learning team. *Management Science*, *33*, 107–123.

Lucas, H. C. (1999). *Information Technology and the Productivity Paradox*. New York: Oxford University Press.

Mackenzie, K. D. (1994). The science of an organization. Part I: A new model of organizational learning. *Human Systems Management*, *13*, 249–258.

March, J. G. (1991). Exploration and exploitation in organizational learning. *Organization Science*, 2, 71–87.

Marshak, R. J. (1998). A discourse on discourse: Redeeming the meaning of talk. In D. Grant, T Keenoy, & C. Oswick (Eds.), *Discourse and Organization* (pp. 65–83). London: Sage.

Marsick, V. J. (1998, October). Individual strategies for organizational learning. Lecture presented at the Workplace Learning Institute, Teachers College, New York.

McCarthy, B. (1999). *Learning Type Measure*. Wauconda, IL: Excel.

McDermott, R. (2000). Why information technology inspired but cannot deliver knowledge management. In E. L. Lesser, M. A. Fontaine, & J. A. Slusher (Eds.), *Knowledge and Communities* (pp. 21–36). Woburn, MA: Butterworth-Heinemann.

McGraw, K. (2009). Improving project success rates with better leadership: Project Smart. Available at www.projectsmart.co.uk/improving-project-success-rateswith-better-leadership.html

Mezirow, J. (1990). *Fostering Critical Reflection in Adulthood: A Guide to Transformative and Emancipatory Learning.* San Francisco, CA: Jossey-Bass.

Milliken, C. (2002). A CRM success story. *Computerworld.* Available at www.computerworld.com/s/article/75730?A_CRM_success-story

Miner, A. S., & Haunschild, P. R. (1995). Population and learning. In B. Staw & L. L. Cummings (Eds.), *Research in Organizational Behavior* (pp. 115–166). Greenwich, CT: JAI Press.

Mintzberg, H. (1987). Crafting strategy. *Harvard Business Review, 65*(4), 72.

Mintzberg, H., & Waters, J. A. (1985). Of strategies, deliberate and emergent. *Strategic Management Journal, 6,* 257–272.

Moon, J. A. (1999). *Reflection in Learning and Professional Development: Theory and Practice.* London: Kogan Page.

Moon, J. A. (2000). *A Handbook for Academics, Students and Professional Development.* London: Kogan Page.

Mossman, A., & Stewart, R. (1988). Self-managed learning in organizations. In M. Pedler, J. Burgoyne, & T. Boydell (Eds.), *Applying Self-Development in Organizations* (pp. 38–57). Englewood Cliffs, NJ: Prentice-Hall.

Mumford, A. (1988). Learning to learn and management self-development. In M. Pedler, J. Burgoyne, & T Boydell (Eds.), *Applying Self-Development in Organizations* (pp. 23–37). Englewood Cliffs, NJ: Prentice-Hall.

Murphy, T. (2002). *Achieving Business Practice from Technology: A Practical Guide for Today's Executive.* New York: Wiley.

Narayan, S. (2015). *Agile IT Organization Design for Digital Transformation and Continuous Delivery.* New York: Addison-Wesely.

Newman, K. S. 1999. *No Shame in My Game: The Working Poor in the Inner City.* New York: Vintage Books.

Nonaka, I. (1994). A dynamic theory of knowledge creation. *Organization Science, 5*(1), 14–37.

Nonaka, I., & Takeuchi, H. (1995). *The Knowledge-Creating Company: How Japanese Companies Create the Dynamics of Innovation.* New York: Oxford University Press.

Olson, G. M., & Olson, J. S. (2000). Distance matters. *Human–Computer Interactions, 15*(1), 139–178.

Olve, N., Petri, C., Roy, J., & Roy, S. (2003). *Making Scorecards Actionable: Balancing Strategy and Control.* New York: Wiley.

O'Sullivan, E. (2001). *Transformative Learning: Educational Vision for the 21st Century.* Toronto: Zed Books.

Peddibhotla, N. B., & Subramani, M. R. (2008). Managing knowledge in virtual communities within organizations. In I. Becerra-Fernandez & D. Leidner (Eds.), *Knowledge Management: An Evolutionary View* (229–247). Armonk, NY: Sharp.

Pedler, M., Burgoyne, J., & Boydell, T. (Eds.). (1988). *Applying self-Development in Organizations.* Englewood Cliffs, NJ: Prentice-Hall.

Peters, T. J., & Waterman, R. H. (1982). *In Search of Excellence: Lessons from America's Best-Run Companies.* New York: Warner Books.

Pietersen, W. (2002). *Reinventing Strategy: Using Strategic Learning to Create and Sustain Breakthrough Performance.* New York: Wiley.

Porter, M. E., & Kramer, M. R. 2011. Creating shared value. *Harvard Business Review, 89*(1/2): 62–77.

Prange, C. (1999). Organizational learning: Desperately seeking theory. In M. Easterby-Smith, J. Burgoyne, & L. Araujo (Eds.), *Organizational Learning and the Learning Organization* (pp. 23–43). London: Sage.

Probst, G., & Büchel, B. (1996). *Organizational Learning: The Competitive Advantage of the Future.* London: Prentice-Hall.

Probst, G., Büchel, B., & Raub, S. (1998). Knowledge as a strategic resource. In G. V. Krogh, J. Roos, & D. Kleine (Eds.), *Knowing in Firms: Understanding, Managing and Measuring Knowledge* (pp. 240–252). London: Sage.

Rassool N. 1999. *Literacy for Sustainable Development in the Age of Information* (Vol. 14). Clevedon: Multilingual Matters Limited.

Reynolds, G. (2007). *Ethics in Information Technology* (2nd edn.). New York: Thomson.

Robertson, S., & Robertson, J. (2012). *Mastering the Requirements Process: Getting Requirements Right* (3rd edn.). Upper Saddle River, NJ: Addison-Wesley.

Sabherwal, R., & Becerra-Fernandez, I. (2005). Integrating specific knowledge: Insights from the Kennedy Space Center. *IEEE Transactions on Engineering Management, 52*, 301–315.

Sampler, J. L. (1996). Exploring the relationship between information technology and organizational structure. In M. J. Earl (Ed.), *Information Management: The Organizational Dimension* (pp. 5–22). New York: Oxford University Press.

Saxena, P., & Jain. R. (2012) Managing career aspirations of generation Y at work place. *International Journal of Advanced Research in Computer Science and Software Engineering, 2*(7), 114–118..

Schein, E. H. (1992). *Organizational Culture and Leadership* (2nd edn.). San Francisco, CA: Jossey-Bass.

Schein, E. H. (1994). The role of the CEO in the management of change: The case of information technology. In T. J. Allen & M. S. Morton (Eds.), *Information Technology and the Corporation* (pp. 325–345). New York: Oxford University Press.

Schlossberg, N. R. (1989). Marginality and mattering: Key issues in building community. *New Directions for Student Services, 48*, 5–15.

Schoenfield, B. S. E. (2015). *Securing Systems: Applied Security Architecture and Threat Models.* Boca Raton, FL: CRC Press.

Schön, D. (1983). *The Reflective Practitioner: How Professionals Think in Action.* New York: Basic Books.

Senge, P. M. (1990). *The Fifth Discipline: The Art and Practice of the Learning Organization.* New York: Currency Doubleday.

Siebel, T. M. (1999). *Cyber Rules: Strategies for Excelling at E-Business.* New York: Doubleday.

Stolterman, E., & Fors, A. C. (2004). Information technology and the good life. *Information Systems Research*, *143*, 687–692.

Swadzba, U. (2010). Work or consumption: Indicators of one's place in the Society. In *Beyond Globalisation: Exploring the Limits of Globalisation in the Regional Context (Conference Preceedings)* (pp. 123–129). Ostrava: University of Astrava Czech Republic.

Swieringa, J., & Wierdsma, A. (1992). *Becoming a Learning Organization, beyond the Learning Curve*. New York: Addison-Wesley.

Szulanski, G., & Amin, K. (2000). Disciplined imagination: Strategy making in uncertain environments. In G. S. Day & P.J. Schoemaker (Eds.), *Wharton on Managing Emerging Technologies* (pp. 187–205). New York: Wiley.

Teece, D. J. (2001). Strategies for managing knowledge assets: The role of firm structure and industrial context. In I. Nonaka & D. Teece (Eds.), *Managing Industrial Knowledge: Creation, Transfer and Utilization* (pp. 125–144). London: Sage.

Teigland, R. (2000). Communities of practice at an Internet firm: Netovation vs. in-time performance. In E. L. Lesser, M. A. Fontaine, & J. A. Slusher (Eds.), *Knowledge and Communities* (pp. 151–178). Woburn, MA: Butterworth-Heinemann.

Tushman, M. L., & Anderson, P. (1986). Technological discontinuities and organizational environments. *Administrative Science Quarterly*, *31*, 439–465.

Tushman, M. L., & Anderson, P. (1997). *Managing Strategic Innovation and Change*. New York: Oxford University Press.

Vince, R. (2002). Organizing reflection. *Management Learning*, *33*(1), 63–78.

Wabike P. 2014. University-Community Engagement: Universities at a crossroad? In P. Wabike & J. van der Linden (eds.), *Education for Social Inclusion* (pp. 131–149). Groningen, the Netherlands: University of Groningen.

Wallemacq, A., & Sims, D. (1998). The struggle with sense. In D. Grant, T. Keenoy, & C. Oswick (Eds.), *Discourse and Organization* (pp. 65–83). London: Sage.

Walsh, J. P. (1995). Managerial and organizational cognition: Notes from a trip down memory lane. *Organizational Science*, *6*, 280–321.

Watkins, K. E., & Marsick, V. J. (1993). *Sculpting the Learning Organization: Lessons in the Art and Science of Systemic Change*. San Francisco, CA: Jossey-Bass.

Watson, T. J. (1995). Rhetoric, discourse and argument in organizational sense making: A reflexive tale. *Organization Studies*, *16*, 805–821.

Wellman, B., Salaff, J., Dimitrova, D., Garton, L. Gulia, M., & Haythornthwaite, C. (2000). Computer networks and social networks: Collaborative work, telework, and virtual community. In E. L. Lesser, M. A. Fontaine, & J. A. Slusher (Eds.), *Knowledge and Communities* (pp. 179–208). Woburn, MA: Butterworth-Heinemann.

Wenger, E. (1998). *Communities of Practice: Learning, Meaning and Identity*. Cambridge, MA: Cambridge University Press.

Wenger, E. (2000). Communities of practice: The key to knowledge strategy. In E. L. Lesser, M. A. Fontaine, & J. A. Slusher (Eds.), *Knowledge and Communities* (pp. 3–20). Woburn, MA: Butterworth-Heinemann.

West, G. W. (1996). Group learning in the workplace. In S. Imel (Ed.), *Learning in Groups: Exploring Fundamental Principles, New Uses, and Emerging Opportunities: New Directions for Adult and Continuing Education* (pp. 51–60). San Francisco, CA: Jossey-Bass.

Westerman, G., Bonnet, D., & McAfee, A. (2014). *Leading Digital: Turning Technology into Business Transformation*. Boston, MA: Harvard Business School Press.

Yorks, L., & Marsick, V. J. (2000). Organizational learning and transformation. In J. Mezirow (Ed.), *Learning as Transformation: Critical Perspectives on a Theory in Progress* (pp. 253–281). San Francisco, CA: Jossey-Bass.

Yourdon, E. (1998). *Rise and Resurrection of the American Programmer* (pp. 253–284). Upper Saddle River, NJ: Prentice Hall.

Index